34th PIANC World Congress 2018

AA006954

Connecting Maritime Hubs Globally

Marinas

Panama City, Panama
7 - 11 May 2018

ISBN: 978-1-5108-8706-0

Printed from e-media with permission by:

Curran Associates, Inc.
57 Morehouse Lane
Red Hook, NY 12571

Printed by Curran Associates, Inc. (2019)

For permission requests, please contact
The World Association for Waterborne Transport Infrastructure (PIANC)
at the address below.

The World Association for Waterborne Transport Infrastructure (PIANC)
General Secretariat
Boulevard Roi Albert II, 20 – B. 3
B-1000 Brussels
Belgium

Phone: +32 2 553 71 61

info@pianc.org

Additional copies of this publication are available from:

Curran Associates, Inc.
57 Morehouse Lane
Red Hook, NY 12571 USA
Phone: 845-758-0400
Fax: 845-758-2633
Email: curran@proceedings.com
Web: www.proceedings.com

TABLE OF CONTENTS

COMMITTEES

INTERNATIONAL ORGANIZING COMMITTEE (IOC)

LOCAL ORGANIZING COMMITTEE (LOC)

Mr. Manuel Benítez, Congress Co-Chair (ACP)
Mr. Rogelio Gordón, Congress Co-Chair (ACP)
Mr. Oscar Vallarino, Local Executive Committee (ACP)
Ms. María Teresa Maure (ACP)
Mr. Ricardo Álvarez (ACP)

INTERNATIONAL SCIENTIFIC COMMITTEE (ISC)

Dr. Philippe Rigo, Vice President of the SC (Belgium)
Dr. Michael Fastenbauer, Chair (Austria)
Mr. Jean Michel Hiver (Belgium)
Mr. Jose Sanchez (USA)
Mr. Wu Peng, Chair (China)
Mr. Jim Stirling, Chair (UK)
Mr. Fabrice Daly, Chair (France)
Mr. Rik Thomas (Belgium)
Mr. Juergen Troegl, Chair (Austria)
Mr. Leonel Temer (Argentina)
Ms. Milou Wolters, Chair (Netherlands)
Mr. Stefan Devocht (Belgium)
Dr. Carsten Thorenz, Chair (Germany)
Dr. Tom De Mulder, Chair (Belgium)
Mr. Holger Schüttrumpf (Germany)
Mr. John Clarkson, Chair (USA)
Mr. Timothy Paulus, Chair (USA)
Dr. Jasna Muskatirovic (Serbia)
Mr. Ian White, Chair (UK)
Mr. Francisco Esteban Lefler, Chair (Spain)
Mr. Ignacio Sanchidrian Vidal, Chair (Spain)
Mr. Hartmut Bruehl, Chair (Germany)

Mr. H.N. Aswath (India)
Mr. Craig Philip, Chair (USA)
Mr. Edward Schmeltz (USA)
Mr. Hans Burcharth (Denmark)
Dr. Mohammadreza Allahyar (Iran)
Ms. Katrina Dodd (Australia)
Mr. Esteban Biondi, Chair (USA)
Mr. Pablo Arecco, Chair (Argentina)
Mr. Nicholas Pansic, Chair (USA)
Dr. Todd Bridges (USA)

LOCAL SCIENTIFIC COMMITTEE (LSC)

Dr. Dimas E. Portillo L., President of the SC (ACP)
Mr. Abdiel Perez (ACP)
Mr. Rigoberto Delgado (ACP)
Mr. Antonio Abrego (ACP)
Mr. Ricardo Álvarez (ACP)
Dr. Ricardo Ungo (ACP)
Mr. Juan Wong, Chair (Panama)

COPYRIGHT AND REPRODUCTION

PREFACE

This book of abstracts contains 284 abstracts presented at the 34th PIANC World Congress held in Panama, Republic of Panama, from 7 to 11 May 2018. PIANC World Congress presents and discusses topics relevant to the waterborne transport and infrastructure sector. It rotates among PIANC's member countries every four years and it is open to members and non-members. Participants, have the opportunity to exchange knowledge and experiences.

PIANC's goal is to advance the sustainable development of shallow and deep-draft navigation issues including dredging and dredged material disposal, navigation and pot infrastructure, recreational navigation and related environmental matters.

The Book of Abstract is arranged according to the main Themes of the Congress namely: Inland Navigation, Dredging, Logistics & Infrastructure, Ports, Marinas, and Environment. Inside each Theme, the abstracts are arranged by the topics of the Theme.

Inland Navigation

- Waterway infrastructures: locks, weirs, riverbanks, others.
- Advances in navigation locks design after the Panama Canal Expansion
- Inland navigation channels: safety and reliability
- Integration of inland waterways into inter-modal supply chain
- Inland navigation, waterways, ports & terminals
- Systems and Infrastructures: design & management for inland navigation projects
- River Information Services (RIS, AIS, others)
- River navigation flow control

- Salt water intrusion
- Maintenance and operation of IW, transport & infrastructures
- Inland navigation in South America

Dredging (in the framework of port and navigation projects)

- Effective planning and execution of dredging projects
- Lessons learned from dredging projects worldwide
- Risk management in dredging projects
- Disposal site planning taking into account alternative land uses
- Current dredging & management innovations

Logistics & Infrastructure

- Challenges for capitalizing navigational channels bordering lands
- Synergies among airports, ports and inter-modal assets
- Integrated management of global supply chains
- Integrating ports and economic special Zones
- Provision of added value logistic services in maritime hubs
- Project management in the Panama Canal Expansion

Ports

- Impact and innovative solutions for hinterland ports
- Dry ports: better and more efficient
- Ports of the future: technologies, automation, traceability
- Port management models
- Maritime Port planning and operations
- Inter-modal connections
- Coastal and Port Engineering (in relation with navigation)
- Storm Surge & Tsunami barriers and Flood protection gates

Marinas

- Sustainable and resilient marina design-Marinas Working with Nature (best practices and case studies)
- Marinas as part of Tourism, Ports and Urban Master Plans (best practices and case studies)
- Planning Framework for Recreational & Tourism Navigation Infrastructure Systems (goals, strategies and challenges)
- Regulatory Framework and Private Marina Development in Countries in Transition (best practices and case studies)
- Regional Cooperation and Standardization of Recreational Navigation Information Systems Transition (best practices and case studies)
- Recreational components of mitigation plans for large navigation infrastructure projects

Environment

- Environmental management in navigation
- Climate change and emissions, energy efficiency, international regulations, carbon markets
- Fresh water availability for operations
- New technologies on infrastructure, pollution prevention, port reception facilities and ballast water
- Multiple purpose water resource systems (transport, energy, recreation, ecosystems, watersheds, potable water, ...)
- Navigation as a green transportation mode
- Societal awareness and responsibility, combining economic growth, environment / sustainability and welfare
- Climate Change Mitigation in the Maritime Industry

All these abstracts represent an invaluable source of information, knowledge, experience, and ideas useful for students, engineers, researchers, scientists, authorities and for everyone interested in the fields covered in this World Congress.

Name________________________________
Title_________________________________
Position______________________________
Affiliation____________________________

MARINAS DESIGN IN AREAS OF HIGHLY ENVIRONMENTALLY SENSITIVE (CASE STUDIES)

Manuel González Moles[1], Ozgur Unay Unay[2], Víctor Jiménez García[3]

ABSTRACT

The calm always comes after the storm and, in our case, after a period of economic circumstances which were unfavorable for us, the sharp improvement in the nautical sports sector has resulted in the need for new berths, whether it be through expansions or new locations. The latter, given the degree of occupancy on the coast as well as the correct environmental protection measures to be applied, are very complicated to place.

As a result, innovation, environment, development, land-use and sustainability must join forces in order to find products and solutions with a similar effect on society, significantly decreasing the environmental impact created, working with natural processes to protect, restore or even improve the environment.

1. GLOBAL DESCRIPTION OF THE WORLDWIDE NAUTICAL OFFER

The recreational nautical industry is an economic activity that has historically been closely linked to both tourism and infrastructure as well as the real-estate sector.

If we take a look at worldwide macro values, according to data from the National Marine Manufacturers Association (hereafter NMMA), we can highlight:

- There are more than 25 million recreational watercraft registered and in service around the world.
- This impacts the service industry by favoring the creation and existence of more than 100,000 small and medium enterprises.
- More than 25,000 marinas of different types and sizes are registered around the world, directly creating one million jobs.

For this reason, the sector must be taken care of given its important impact. However, at the same time, we need to monitor and condition its continuity and growth while fully respecting the basic environmental principals, assuring the future sustainability of the environments in which both marinas and economically associated businesses are found.

Today, a true Ecosystem (a name also used by the NMMA) has been created around the maritime sector which encompasses numerous sectors and activities. This ecosystem is a true economic force creating value that cannot be jeopardized and must be strengthened by the sensible use of natural resources both during design and when developing facilities.

This important ecosystem is made up of retail stores, restoration, marina infrastructure, vessels, sailors, accessory factories and distribution, sailing schools, distributors, service providers, brokers and many more.

As far as we understand, recreational nautical infrastructure has often been linked to a “payback” resulting from real-estate investment, forgetting that, in this sector, a significant number of activities intertwine which, in addition to adding considerable value, help repay the investment necessary to create the infrastructure itself.

Additionally, outsiders (as well as some insiders) associate the nautical industry with giant yachts, luxury cars and products which are completely inaccessible to the middle class, which should be the real target population.

---

[1] MSc. Civil Engineer, CEO, UG21 Consultores de Ingeniería S.L., mgonzalez@ug21.com
[2] MSc. Civil Engineer, President, UG21 Consultores de Ingeniería S.L., ozgurunay@ug21.com
[3] MSc. Civil Engineer, PMP®, Project Engineer, UG21 Consultores de Ingeniería S.L., vjimenez@ug21.com

All data points to recreational boaters being members of the middle class, with moderate income, who see this pastime as one more target within today's day-to-day value scale.

According to data from the NMMA, 95% of watercraft registered in the United States measure 8 meters (24 feet) or less in length. We must also add that only 1% of boats registered exceed 12 meters (37 feet). This figure is also reflected in other countries such as:

- France: 91% of boats measure 8 meters or less in length.
- Queensland and New South Wales: 94% of boats measure 8 meters or less in length.

Economic models also reflect these figures.

At a recent meeting of the International Council of Marine Industry Associations (ICOMIA) in Amsterdam, Adjiedj Bakas also pointed out that the recreational boating industry was on the path to a collaborative economy which presumes a boom and revolution we must be ready for. Bakas stated:

..."*the middle class will have less money but more time thanks to robotization. In the* ***Boating and Marina*** *market trends, people will look for affordable boats with low maintenance costs, often sharing them through* ***collaborative economy****. Marinas will go from being parking lots for boats to places where people share a* ***community lifestyle***".

..." *We have to change the garage into a* ***leisure destination***"...

However, we cannot forget about the mega yacht market which, in spite of the economic crisis we've seen, has not stopped growing. According to Gregorio Méndez de la Muela at the V Foro Náutico de Colombia, held in Santa Marta, *"In spite of the economic uncertainty after the 2007 economic crisis, the* ***mega yacht market continues to grow annually****. In recent years, shipyards in Italy, France and the United Kingdom have built nearly 500 yachts."*

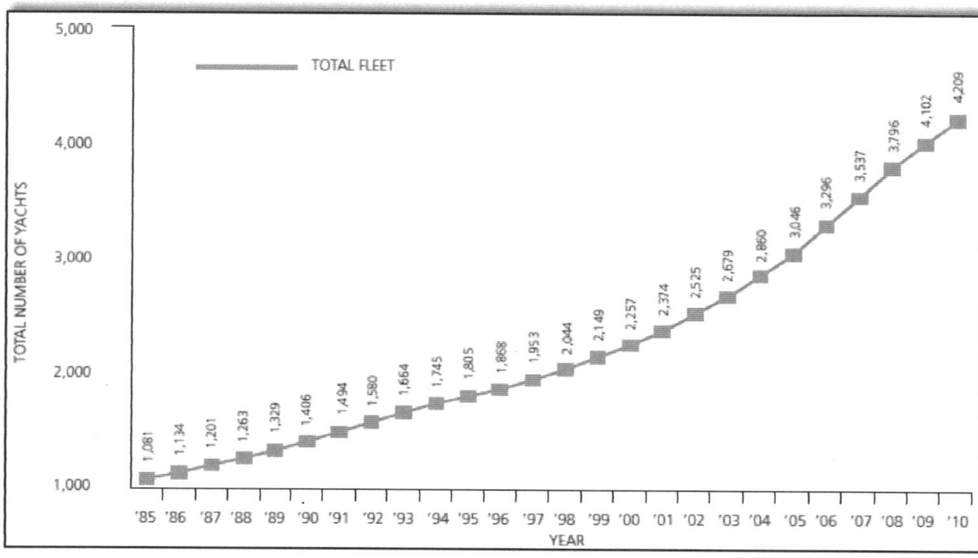

Figure 1: Evolution of the Mega Yacht fleet since 1995. (V Foro Náutico Colombia, 2017)

In summary, the recreational nautical industry is booming and all signs point to significant future growth which we must be prepared for in order to create new infrastructure able to meet the growing demand while having as small an impact as possible.

Our contribution to this prestigious conference will discuss this: the possibility of balancing highly environmentally sensitive areas with the construction of low-impact marina facilities, controlled investment and highly beneficial effects for the local economy.

We will mention three practical cases, some already in place, which show major investment is not necessary and that the return on these facilities is guaranteed without the need for large real-estate promotion or land creation that is incompatible with coastal sustainability.

2. NEW FACILITIES: ENVIRONMENTAL CONDITIONS AND RISKS

There is a mature market in developing countries, where the majority of concessions are advanced. The mooring offer adapted to demand until 2007, but since then there has been practically no evolution.

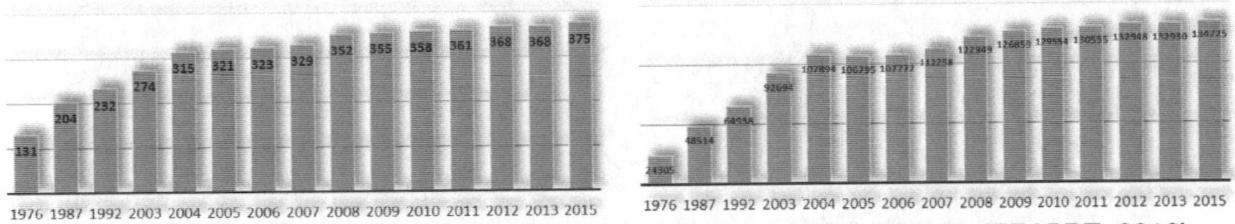

Figure 2: Evolution of marinas (left) and moorings (right) in Spain. (FEAPDT, 2016)

The number of vessels has continued to grow while the number of berths available has not, thus resulting in today's saturation. This shows the difficulties arising from expanding or creating new marina infrastructure. A marina is not simply a place to tie up boats.

Marinas are facilities designed by specialized engineers to be able to resist extreme wave heights, currents and winds, along with actions caused by boats and humans. This entails significant design of protection structures if the waters are not protected naturally. At the same time, the impact on the local hydrodynamics, coastline dynamics and marine environment must be taken into account.

Furthermore, economic, market and environmental issues must be integrated and we must calculate whether investment will be worthwhile, what the current and future nautical trends are and if the market will be sustained in the long term. The high depreciation costs for protective structures are not compatible with concession periods (a maximum of 30 years in Spain), meaning the payback on investment is not guaranteed. This issue is magnified by the correct and necessary environmental restrictions of the *Ley de Suelo* (Land Law), which prohibits real-estate development being associated with the marina, as was done in the past. Puerto Sotogrande, Puerto Banús, Marina del Este, etc. are worth mentioning as large investments that, without taking real-estate development into account, would have been practically impossible to have been repaid.

Figure 3: Marinas Sotogrande, Banús y Marina del Este

As a result, <u>building new facilities</u> in developed countries is a very complex task. In Andalusia, for example, where its coastline is its greatest strength, not one marina has been built in unprotected water in the past 20 years.

Another alternative is to <u>expand existing ports</u>, which, in any case, also has major obstacles to success. The previously mentioned concession periods limit the large investment necessary for expansion in the case of unprotected waters. This is the case of the Puerto Deportivo in Benalmádena and Puerto José Banús as well as the recently failed concession of the Marina la Bajadilla.

Figure 4: Marina la Bajadilla present and future. (Berenguer Ingenieros, 2012)

Yet another alternative or solution is to analyze the coastline in order to determine those zones which have their own natural protective condition, making it unnecessary to build rigid protective constructions that can adversely affect the costal dynamic, thus making them opportune areas to place new maritime facilities. Working with, not against, natural processes can lead to less costly and more sustainable solutions.

If opportune measures are taken, these areas, initially dismissed for their high environmental sensitivity, could harbor a marina with absolutely no environmental effects. Such sensitive areas require measures which allow us to reduce the effect on marine ecosystems.

Biotic, abiotic and socioeconomic environmental components must be taken into account when evaluating the environmental impact of each stage in marina design and construction.

Measures must be arranged to control solid, liquid and gas emissions that may alter surroundings. In general terms, control measures that must be taken include:

- Air Quality: Occasional irrigation, using tarps on trucks, using vehicles with emission control systems, restricting working hours to avoid disruption during rest periods, etc.
- Soil Quality: Heavy equipment or areas with impervious containment, promoting suitable classification and disposal of waste or solid urban waste, etc.
- Protection of Flora and Fauna: Controlling periods of maximum wildlife activity, promoting activities to protect the ichthyofauna present, beacons to minimize the area of operation, etc.
- Water Quality: Using geotextile barriers, promoting an emergency plan in case of unexpected spillage, promoting an annual code of conduct manual for users, etc.

In addition to designing a structure that is compatible with nature, it is fundamental for structures to include the equipment and facilities necessary to reduce possible environmental effects. Among others, this must include:

- Implementing systems that use renewable energy to supply electricity.
- Using energy-efficient luminaires.
- Bilge cleaning systems.
- Removable anti-contamination barrier to avoid the escape or entry of spills in case of accidental spillage and to facilitate marina cleaning.

At the same time, there is currently no European regulation regarding ecological requirements for watercraft, but it is in the development stage, thus the use of hybrid and ecological watercraft will be another aspect to take into consideration.

By keeping the abovementioned practices in mind, we can be sure to have modern and complete facilities for recreational nautical use and complementary activities, without any risk for a location which is invaluable from the environmental perspective.

3. PROTECTIVE CONDITIONS AND OPTIMAL WAVES

In recent years, we have had the opportunity to design several marinas in specially protected areas. Particularly noteworthy are the nautical sports facilities in the Río Piedras natural marshlands and the El Rompido sandbar in Huelva, Spain, the Archipelago of San Andrés, Providencia and Santa Catalina in the Colombian Caribbean, and the Topocoro reservoir in Santander, Colombia, all of which are classified as highly environmentally sensitive.

Occasionally, nature does us a favor by offering naturally protected areas which do not require rigid protective constructions that can adversely affect the costal dynamic, thus leading to collateral damage. Such is the case in the areas mentioned above. In the first case, we have the El Rompido sandbar to provide refuge for watercraft, a coastal reef in the second case, and the Topocoro marina is located in a reservoir.

The protection provided by the area itself allows for "permeable" structures, like a floating breakwater, to guarantee marina operability. Energy dissipation is left to the natural structures, leaving short-period residual waves to the floating barriers.

The following are the required wave conditions in marinas:

- Hs = 0.25-0.30 m ....Comfort limit for small watercraft (L < 12 m)
- Hs = 0.35-0.40 m ....Comfort limit for large watercraft (L > 20 m)
- Hs = 0.50 m ....Guaranteed limit for floating docks
- Hs = 0.60 m ....Safety limit for small watercraft (L < 12 m)
- Hs = 0.80 m ....Safety limit for large watercraft (L < 20 m)

The aforementioned concrete docks or floating breakwaters provide extraordinary attenuating effects for low-period swells characteristic of semi-protected areas.

Figure 5: View of floating breakwater attenuating effect.

Floating breakwaters have traditionally been believed to be unstable structures associated with small recreational vessels. This is due to the fact that, until recently, the preferred construction material was metal profiles (aluminum or hot galvanized steel).

The boom in concrete as a construction element in void boxes completely changes the scene and, for the first time, floating barriers can be used which feature stability and resistance similar to that of fixed docks made with slabs of the same material. Additionally, both maintenance and environmental impact are negligible.

By forming a single segment without individual floats, concrete wave barriers offer great stability (from 400 to 1,200 Kg/m^2) and resist waves in semi-protected areas quite well (up to 1.5 m high waves).

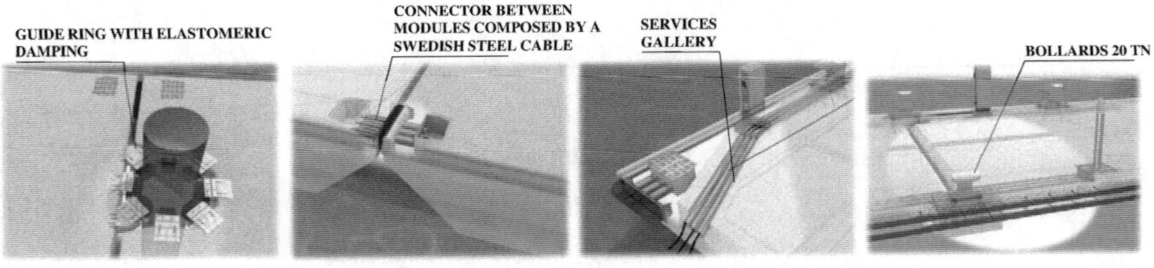

Figure 6: Floating breakwater details (MSI)

The units have service galleries through which the networks needed for water supply and electricity run. They are anchored with chains and deadman anchors or piles. However, in the case of the latter, it should be noted that studies are scarce. By using piled wave barriers, induced movements are eliminated, only allowing degree of freedom z in their movement. This type of equipment offers two advantages:

a) It doesn't lead to any snaking on the ground. Some manufacturers also avoid this by installing post-tensioned cable.
b) It allows mooring larger watercraft on its sheltered coast (yachts and mega yachts).

Floating breakwaters can be used in both protected and semi-protected areas which require high quality and stability.

Their dissipating effect is backed by the reduced model from the Institute of Applied Hydrodynamics (INHA, Spanish acronym) in its study "Three-dimensional tests of floating docks to analyze their hydrodynamic behavior against wave transmission which our firm has had access to"[4] at the request of Marina System.

Figure 7: View of INHA's reduced model.

Given the fact that a large number of phenomena intervene in the interaction between waves and the structure, there is no floating breakwaters that is a completely reflective system. To estimate transmission coefficient k_T of the structure, we need to study the variables that influence it. We must keep in mind the incident wave, the geometry of the wave barrier, its mechanical properties, moors available and surrounding water viscosity. Taking all these basic parameters into consideration, the dimensional analysis provides expressions like:

$$k_T = f\left(\underbrace{\frac{d}{L}, \frac{H}{L}}_{\text{Waves}}, \underbrace{\frac{B}{L}, \frac{h}{d}}_{\text{Geometry}}, \underbrace{\frac{M}{\rho B h}, \frac{I}{MB^2}}_{\text{Mass}}, \underbrace{\frac{h_G}{h}, \frac{kB}{Mg}}_{\text{Berths}}, \theta, \underbrace{\frac{B\sqrt{gd}}{\gamma}}_{\text{Viscosity}} \right)$$

The most important parameters are considered to be that related to depth (d/L), that which controls the vertical distribution of energy flow, and that related to width (B/L). Numerous studies have obtained k_T curves solely based on said parameters.

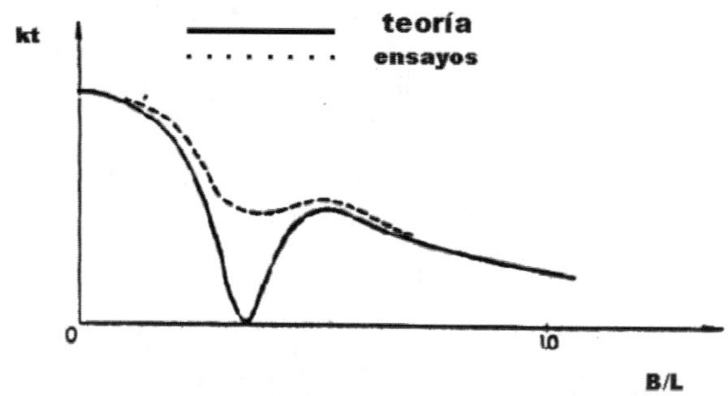

Figure 8: Typical attenuation curves.

[4] Original title: "Ensayos tridimensionales de pantalanes flotantes para analizar su comportamiento hidrodinámico frente a la transmisión de oleaje, al que nuestra firma ha tenido acceso"

Floating breakwaters are structures whose movements are induced by incident waves. These movements, similar to those of a mechanical system undergoing a sinusoidal force, play an integral role in determining the strain on moors and wave attenuation.

For regular waves, there is relatively good agreement between experimental results and theory, with the exception of resonance, where radiated and refracted waves theoretically have opposite phases and the k_T should thus be equal to zero. This is due to the fact that as soon as the dock movements become predominant, the linear model is no longer valid. This is also true because a purely regular wave and perfectly elastic moors do not exist.

In summary, concrete wave barriers perform best when the relationship between ω/ρ (angular wave frequency/angular system frequency) exceeds the unit. The movement amplification factor decreases and the phase approaches 180º. This occurs in short-period waves where the movements of the structure and the waves are significantly lagged.

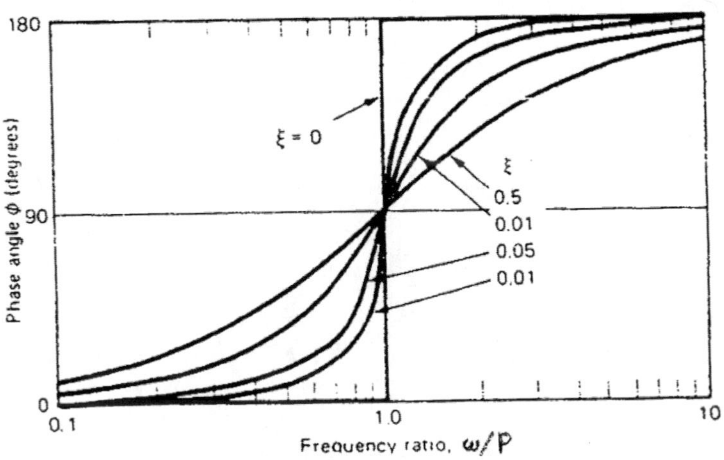

Figure 9: Phase relationship.

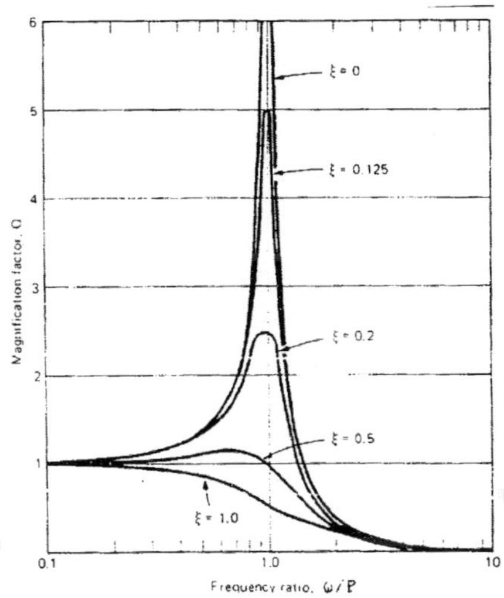

Figure 10: Magnification factor.

We must thus keep in mind that for a specific wavelength, the transmission coefficient k_T is practically independent with regard to the height of the incident wave, except for periods near natural periods of dock oscillation (roll and heave), in which the system produces wave resonance phenomena resulting in increases (when in phase) or decreases (before gaps) in the transmission coefficient according to wave height.

The most noteworthy results of the INHA's reduced model are:

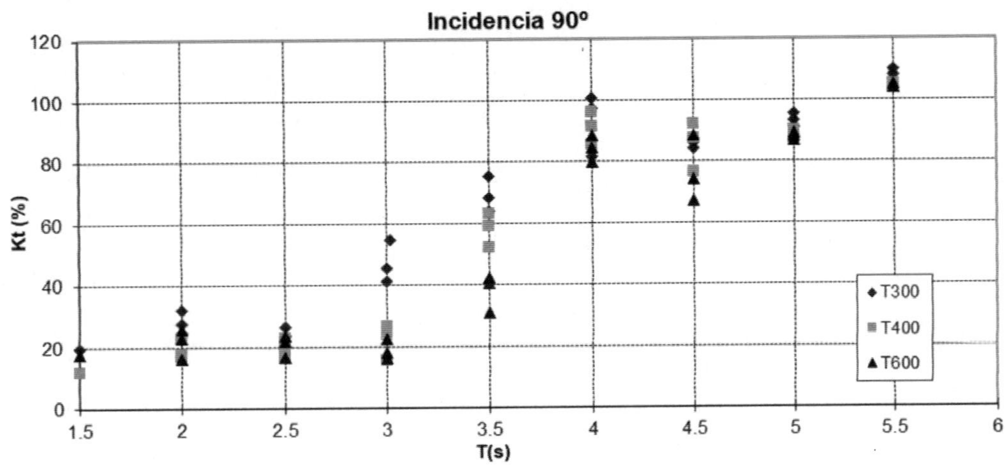

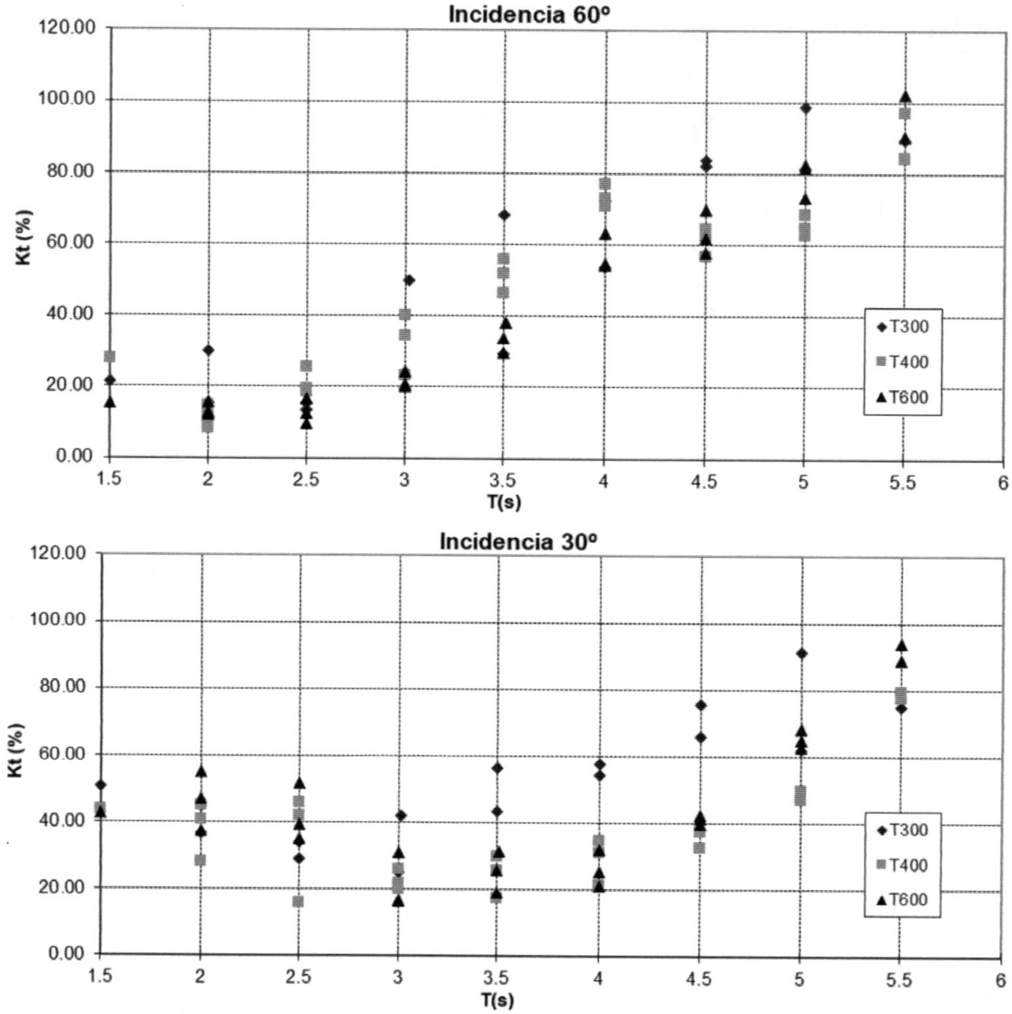

Figure 11: Transmission coefficient for periods between 1,5 and 5,5 s. (MSI and INHA)

In the table above, we can note that the floating docks are able to provide significant attenuation for swells with periods up to 4 seconds. Periods that exceed 5 seconds require excessively wide structures or highly innovative designs.

4. CASE ESTUDIES

We would like to share the following case estudies:

- A.D.N. Nuevo Portil nautical sports facilities, T.M. Cartaya (Huelva, Spain)
- Asociación Náutica San Miguel nautical sports facilities, T.M. Cartaya (Huelva, Spain)
- Club náutico de Rio Piedras nautical sports facilities, Punta de la Barreta, T.M. Cartaya (Huelva, Spain)
- Marina for yachts and sailboats on the island of San Andrés, Archipelago Department of San Andrés, Providencia and Santa Catalina (Colombia)
- Marina in Topocoro Reservoir, Department of Santander, Colombia

These marinas consist of approximately 400 moorings designed in Ría del Piedras and 160 spaces planned for the marinas in Colombia. They consist of sections of piled floating jetty, protected by a floating concrete breakwater anchored to the ocean floor.

These are ambitious marinas designed for customer use and enjoyment, while at the same time respecting the environment.

In addition to a structural design that is compatible with nature, it is essential that the marinas feature fixtures and facilities necessary to reduce possible environmental impacts.

MARINAS AT THE MOUTH OF THE RÍO PIEDRAS, HUELVA (SPAIN)

An area has been chosen within the Río Piedras estuary classified as L.I.C. (Place of Community Interest, Spanish acronym) and close to the Río Piedras natural marshlands and El Rompido sandbar. In order to avoid negative effects in the area, important environmental measures have been taken and navigation and maneuvering have been improved in the estuary by eliminating, or at least minimizing, indiscriminate mooring.

Figure 12: Mouth of the Río Piedras year 2011 and year 2016

PIEDRAS ESTUARY CONSTRAINTS

The El Rompido sandbar, which forms a part of the Río Piedras natural marshlands and the El Rompido sandbar, protects the area dedicated to harboring marinas built in the Piedras estuary.

The waves we find in these areas are localized waves created by the wind or the passing of watercraft. With appropriate regulation of maximum speeds through the canal, the waves created can easily be attenuated by the floating concrete barriers proposed to protect these areas.

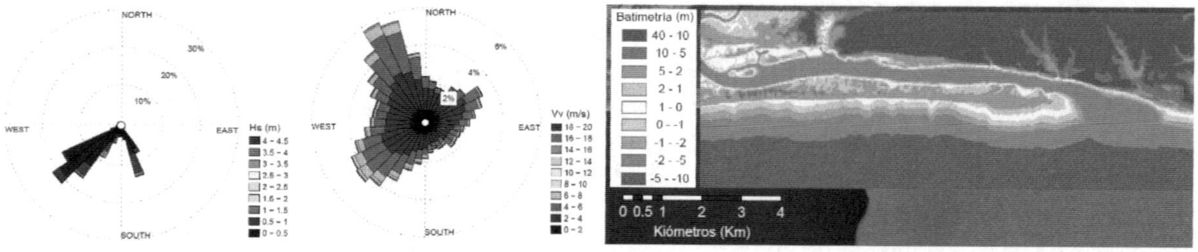

Figure 13: Piedras estuary constraints

A.D.N. NUEVO PORTIL NAUTICAL SPORTS FACILITIES, T.M. CARTAYA (HUELVA, SPAIN)

The following are the characteristics of the proposed marina:

- Total built-up area: .... 5.337,36 m²
- Number of moors: ....387 units
- Watercraft length: .... 8, 10 y 12 m

Protection works

- Concrete floating breakwater: ....304,25 m + 97,24 m + 145,95 m

Docks

- Floating units ....12 x 3 m y 12 x 2,5 m
- Number of fingers ....179 units

Walkway to land

- Fixed walkway: .... 101.05 m

Figure 14: A.D.N. Nuevo Portil nautical sports facilities, T.M. Cartaya (Huelva, Spain)

ASOCIACIÓN NÁUTICA SAN MIGUEL NAUTICAL SPORTS FACILITIES, T.M. CARTAYA (HUELVA, SPAIN)

The following are the characteristics of the proposed marina:

- Total built-up area: .... 16.696,76 m^2
- Dredging: 46.672,25 m3
- Number of moors: .... 379 units
- Watercraft length: .... 8, 10 y 12 m

Protection works

- Concrete floating breakwater: ....255,00 m + 133,50 m + 145,65 m

Docks

- Floating units .... 12 x 3 m y 12 x 2,5 m
- Number of fingers .... 171 units

Walkway to land

- Fixed walkway: .... 73,92 m

Figure 15: Asociación Náutica San Miguel nautical sports facilities, T.M. Cartaya (Huelva, Spain)

CLUB NÁUTICO DE RIO PIEDRAS NAUTICAL SPORTS FACILITIES, PUNTA DE LA BARRETA, T.M. CARTAYA (HUELVA, SPAIN)

The following are the characteristics of the proposed marina:

- Total built-up area:: .... 15.680,67 m²
- Number of moors: ....411 units
- Sea walls: 1.693,90 m3
- Watercraft length: .... 8, 10 y 12 m

Protection works

- Concrete floating breakwater: ....257,00 m + 133,00 m + 157,29 m

Docks

- Floating units ....12 x 3 m y 12 x 2,5 m
- Number of fingers ....197 units

Walkway to land

- Fixed walkway: .... 76,72 m

BEFORE ◄ ┬ ► NOW

Figure 16: Club náutico de Rio Piedras nautical sports facilities, Punta de la Barreta, T.M. Cartaya (Huelva, Spain)

MARINA FOR YACHTS AND SAILBOATS ON THE ISLAND OF SAN ANDRÉS, ARCHIPELAGO DEPARTMENT OF SAN ANDRÉS, PROVIDENCIA AND SANTA CATALINA (COLOMBIA)

The Archipelago of San Andrés, Providencia and Santa Catalina (Colombia) forms a part of the World Network of Biosphere Reserves with the name SEAFLOWER. All of the necessary measures were taken to obtain an environmental license according to the environmental authorities at the national and regional level such as the National Authority on Environmental Licenses (ANLA, Spanish acronym) and the Corporation for the Sustainable Development of the Archipelago Department of San Andrés, Providencia and Santa Catalina – Coralina.

Figure 17: Archipelago Department of San Andrés, Providencia and Santa Catalina (Colombia), SEAFLOWER

San Andrés Island Constraints

The area is naturally protected from external ocean conditions thanks to its favorable topographical characteristics, among which we can highlight the attenuating effect of the coral barrier that is near the Eastern coast of the island. This barrier absorbs the waves that reach the coastline and reflects those with longer wavelengths and allows for the propagation of smaller ones.

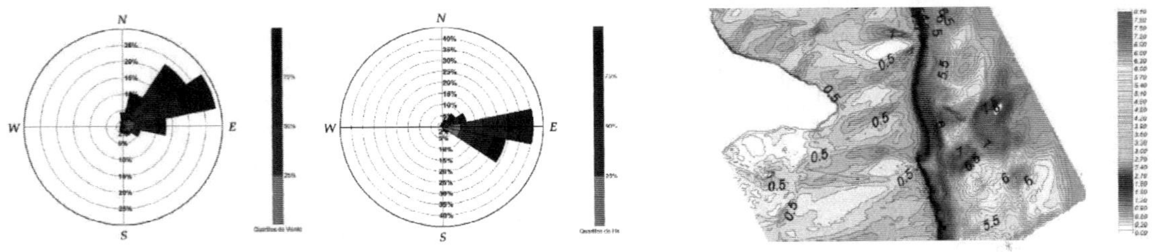

Figure 18: San Andrés Island Constraints

Proposed solution

The following are the characteristics of the proposed marina:

- Total built-up area:.. 1334.26 m²
- Number of moors:..155 units
- Watercraft length: .... 8, 10, 12, 15, 20 y 30 m

Protection works

- Concrete floating breakwater: ....104,29 m + 96,29 m + 109,27 m

Docks

- Floating units .... 12 x 2,5 m
- Number of fingers ....72 units

Walkway to land

- Fixed walkway: .... 110 m

Figure 19: Marina for yachts and sailboats on the Island of San Andrés

Other difficulties: Given the occupancy rate of the land in the area, it was necessary to design a multi-purpose octagonal platform with a surface area of 1090 m^2.

Considering basic user needs in the platform area, a program was developed which includes all uses of the marina (control, administration, sanitation facilities, etc.). It offers a series of commercial uses such as a mini-market, café and retail areas. Finally, in line with ecological considerations and preserving biodiversity in the area, a multi-purpose room for training, a garden with natural coral using Biorock technology and a storage and maintenance hangar for non-motorized watercraft have been proposed.

Figure 20: Multi-purpose area San Andrés Marina

MARINA IN TOPOCORO RESERVOIR, DEPARTMENT OF SANTANDER, COLOMBIA

The Topocoro reservoir, created after the construction of the Sogamoso Hydroelectric Plant in 2014, has become the largest in the country with a capacity of 4.8 billion cubic meters.

The location chosen for the marina is within the integrated management district Serranía de los Yariguies, inside the 100-meter buffer zone around the park. As a result, the National Authority on Environmental Licenses (ANLA, Spanish acronym) and the Regional Autonomous Corporation of Santander (CAS, Spanish acronym) have been contacted in order to take their requirements into account throughout the development of the project.

Topocoro Reservoir Constraints

The greatest constraint considered in this design is the variation in water level within the reservoir. Levels vary up to 40 meters annually, making it necessary to consider a maximum variation of 60 meters in the design.

The main external agent to take into account is the wind, as well as the waves it creates.

Proposed solution

To adapt to the large variation in water level that must be considered in the design, the construction of an amphibious walkway consisting of aluminum units with polystyrene floats is planned. This structure is supported by a piled steel frame along the length of the existing slope, making its maximum gradient 20%, requiring the length of the walkway to be longer than 300 meters.

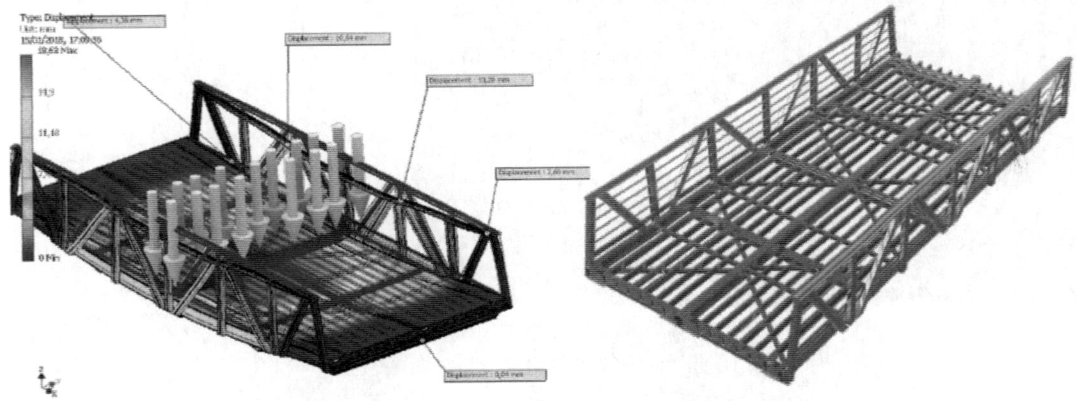

Figure 21: Renders of projected modules

Units are connected using reinforced EPDM elastomer seals measuring 150x150x48 mm which have been reinforced with stainless sheet metal. Four units of pure steel cable will be used throughout the length of the walkway as a safety element. It is designed to avoid being tensioned, but tight. The units are supported by piled horizontal members.

The dock area is protected from nearby waves by a wave barrier consisting of floating concrete docks attached to the ground with cables driven by hybrid winches.

These winches assure that the walkway rests along the planned axis and fix the dock area in position.

The driving of the winches is planned to work in three ways:

- Using a **command panel** located in the floating area. The marina operator can pull or release the winch cable as needed by pressing buttons. This signal is sent via cable.
- Winches can also be driven by **radio control**. A radio control can drive up to 12 winches at a time.
- Finally, they can be driven **manually**. A multiplier will be installed which will allow increasing the torque of submerged cable.

The following are the characteristics of the proposed marina:

- Total built-up area:.. 6.424,50 m²
- Number of moors:...163 units
- Watercraft length: .... 8, 10, 13 y 20 m

Protection works

- Concrete floating breakwater: .... 70 m + 90 m + 90 m

Docks

- Floating units .... 12 x 2,5 m, 12 x 3 m y 12 x 5 m
- Number of fingers ....74 units

Walkway to land

- Fixed walkway: .... 320 m

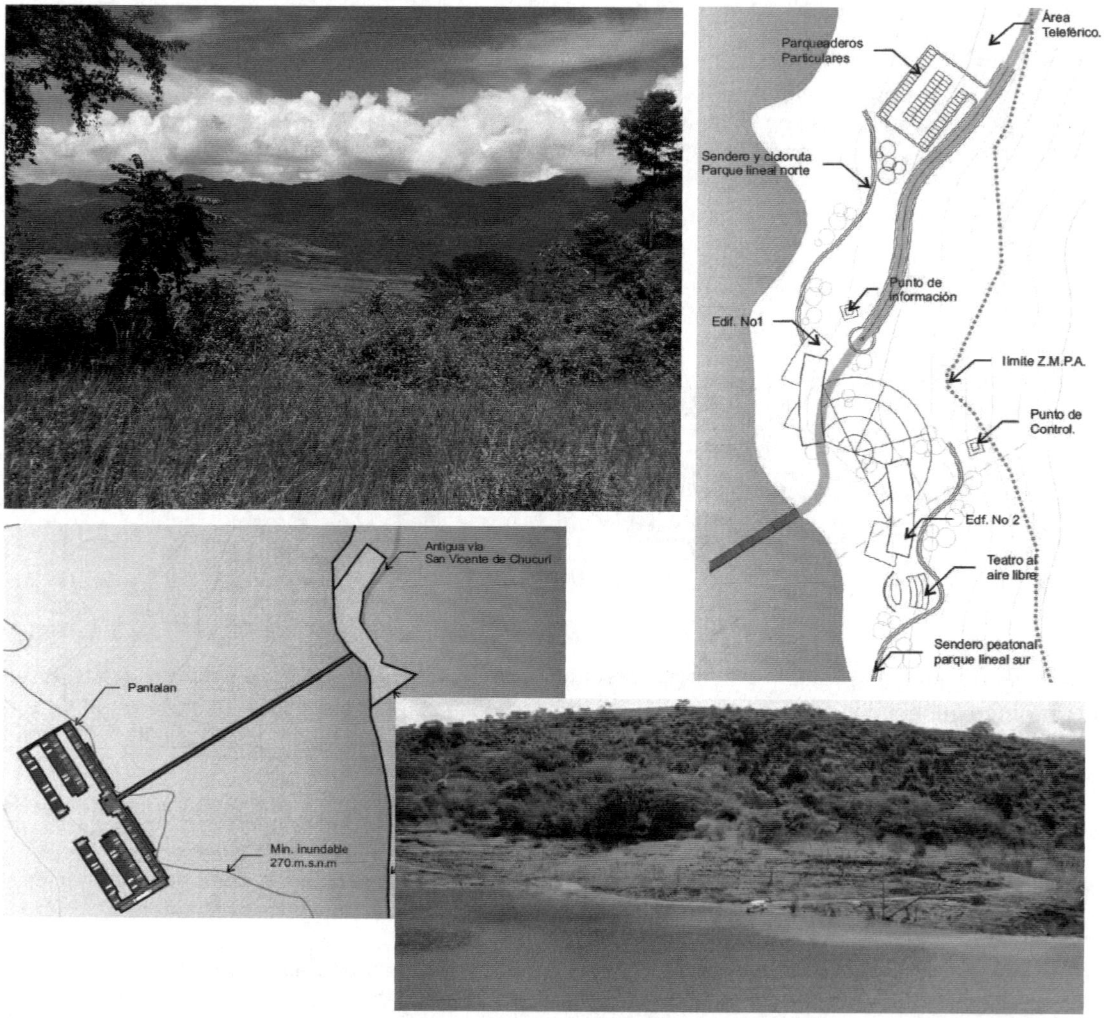

Figure 22: Marina in Topocoro reservoir, Department of Santander, Colombia.

Conclusions

In summary, nautical activity does not have to oppose environmental conservation and suspected water pollution in the area where activity takes place. An example of sustainable activity can be set if it is properly regulated, designed and controlled.

Bibliographical References

- Asociación Nacional de Empresas Náuticas (A.N.E.N.)
- Bakas, A., The International Council of Marine Industry Associations (I.C.O.M.I.A.) -World Marinas Conference, Amsterdam, 2016.
- Berenguer Ingenieros S.L. image of the project of a new Marina of Marbella. https://berengueringenieros.com
- Doménech, J.L. y Menéndez R. (2004) "El papel de los puertos marítimos en la conservación de la biodiversidad y en la gestión integrada del litoral", VII CONGRESO NACIONAL DEL MEDIO AMBIENTE
- Federación Española de Asociaciones de Puertos Deportivos y Turísticos (F.E.A.P.D.T.)
- Marina System Ibérica (M.S.I.), Ensayos tridimensionales de Pantalanes flotantes para analizar su comportamiento hidrodinámico frente a la transmisión de oleaje. Instituto de Hidrodinámica Aplicada (I.N.H.A.)
- Méndez de la Muela, Gregorio, "La Náutica en Ciudades Turísticas", V FORO DE TURISMO NÁUTICO COLOMBIANO, Santa Marta, 2017.
- National Marine Manufacturers Association (N.M.M.A.), "The global Boating Market", 2016.

VALUE OF 3D PHYSICAL MODELING IN HARBOR DESIGN - GATEWAY HARBOR CHICAGO CASE STUDY

Andrew Cornett[1], Scott Baker[2], Bill Weaver[3]

ABSTRACT

In this paper the important role that physical modeling can play in enabling the efficient and optimized design of ports, harbors, and marinas is discussed with reference to a specific project example, known as Gateway Harbor, Chicago. A 3D hydraulic model study was commissioned to help optimize and validate the design of Gateway Harbor, a new harbor proposed for a site beside Navy Pier on the shore of Lake Michigan near the center of Chicago, Illinois, USA.

A 3D physical model of the eastern/outer portion of the proposed harbor was constructed in a 36m x 30m multidirectional wave basin at a geometric scale of 1/30. Accurate reproductions of a range of extreme wave conditions were generated in the model by means of a sophisticated 60-segment directional wave generator. The model was fitted with instrumentation to measure wave conditions within the harbor and uplift pressures on an existing deck-on-pile structure running along the south side of Navy Pier. After establishing wave conditions in the new harbor and wave-induced uplift pressures on the pile-supported deck structure for existing conditions and for the baseline harbor layout, the focus shifted to investigating alternative harbor layouts that would alleviate the uplift pressures and reduce wave agitation and wave overtopping as much as possible without increasing project costs substantially. Over twenty alternative harbor layouts were modeled and assessed in the physical model study. Some of the more effective harbor layouts were able to reduce uplift pressures, agitation levels and overtopping considerably compared to the baseline layout. The finding that excessive uplift pressures could be reduced to acceptable levels by making relatively small and inexpensive changes to the harbor layout was an important factor in the viability of the Gateway Harbor project. This important refinement and optimization of the baseline design would not have been possible without the 3D physical model study.

INTRODUCTION

In this paper the important role that physical modeling can play in supporting the efficient and optimized design of ports, harbors, and marinas will be discussed. While the power and capability of numerical modeling approaches has increased dramatically in recent decades, some important gaps remain where physical modeling approaches can deliver better, more reliable answers. Physical modeling facilities and technologies have also improved in recent decades, and physical modeling studies, particularly those conducted at large scale, remain the preferred approach to optimize the layout and design of rubble-mound breakwaters and revetments to suit local conditions, and validate the performance of proposed designs prior to construction. Physical model studies also remain the best approach to predict the magnitude and character of wave-induced uplift pressures on the underside of pile-supported deck structures, and to develop and evaluate alternative strategies for attenuating uplift pressures. Physical model studies are also the best approach for predicting overtopping flows due to waves at complex three dimensional structures, and predicting the behavior of ships moored within harbors. All of the important physical processes governing wave propagation and wave structure interactions, such as wave refraction, diffraction, reflection, wave breaking, non-linear wave-wave interactions, wave run-up, overtopping, interstitial flows, and armor unit stability, can be reproduced in a realistic manner in a large-scale 3D physical model.

The important role that a 3D physical model study can play in supporting the design of a new harbor will be illustrated through reference to a specific project. Gateway Harbor has been proposed as a new harbor to be constructed beside Navy Pier on the shore of Lake Michigan near the center of the

[1] Program Leader, Ocean, Coastal & River Engineering, National Research Council Canada, Ottawa, Canada

[2] Research Engineer, Ocean, Coastal & River Engineering, National Research Council Canada, Ottawa, Canada

[3] Vice President, AECOM, Chicago, IL, USA

City of Chicago, Illinois, USA. Gateway Harbor will be located just south of Navy Pier, and will offer sheltered moorage mainly for recreational yachts, tour boats, and passenger ferries. The project site is located behind an outer breakwater which offers partial sheltering during storms. However, because the outer breakwater is a low-crested structure that experiences a large amount of overtopping during extreme events, the Gateway Harbor site is exposed to moderate wave action during design events with elevated water levels, large waves, and strong onshore winds.

A 3D physical model study was commissioned to support the design of the new harbor. The model study was commissioned by AECOM, funded by the City of Chicago, and conducted by the Ocean, Coastal and River Engineering research centre (OCRE) of the National Research Council, Canada (NRC). Key issues to be addressed included:

a) Wave agitation within the new harbor, which is strongly influenced by the wave overtopping passing over the low-crested outer breakwater.
b) Wave uplift pressures on the underside of a lengthy existing pile-supported deck structure.
c) Potential optimizations to the harbor layout and the design of the new structures to reduce wave disturbance and wave loads, reduce costs, and improve constructability.

A three-dimensional 1/30 scale physical model of the proposed harbor was constructed in NRC-OCRE's 36m by 30m directional wave basin located in Ottawa, Canada. Accurate reproductions of a range of extreme wave conditions were generated in the model by means of a sophisticated 60-segment directional wave generator. The model was fitted with instrumentation to measure wave agitation within the harbor and uplift pressures on the existing deck-on-pile structure running along the south side of Navy Pier, and with several video cameras to monitor conditions in the model.

An initial series of tests was conducted to verify the incident wave conditions in the model for existing conditions without the new harbor. Following this, a 1/30 scale model of the eastern/outer part of the new harbor was constructed and tested in a wide range of extreme water levels and wave conditions particular to the site. The new breakwater structures were constructed using rock materials that were selected to reproduce the hydraulic performance and submerged stability of the materials specified in the prototype design. More than twenty alternative harbor layouts were simulated in the model, and the results of these studies have been assessed to help develop an optimized, cost-effective design which minimizes wave uplift forces, wave agitation, and construction costs. After modifying the model to simulate the inner (western) part of the proposed harbor, a final series of tests was performed to investigate wave agitation throughout the inner harbor.

The study results showed that the initial harbor layout generated significant uplift pressures on the deck-on-pile structure running along the south side of Navy Pier under certain extreme wave conditions and water levels. Numerous modifications to mitigate the issue were subsequently developed, assessed, and verified in the model in an efficient and cost-effective manner. The most effective layouts delivered substantial and important reductions in wave loads, wave disturbance, and wave overtopping that were an important factor contributing to the viability of the new harbor. Such extensive and effective optimization could not have taken place without the 3D physical model study.

Many researchers have studied wave-in-deck loads over the past decade using a variety of theoretical, experimental, and numerical methods. Wave-in-deck loads on various pile-supported coastal structures such as jetties, piers, wharves and bridges have been studied experimentally by Cornett et al. (2013), Tirindelli et al. (2003), Cuomo et al. (2007, 2009), Murali et al. (2009), and Meng et al. (2010). All these authors analyzed data from scale model tests to investigate the pressures and loads on beam and deck elements subject to wave impact under various conditions.

Project Background

Gateway Harbor will be located on the western shore of Lake Michigan near the center of the City of Chicago, between Navy Pier and the entrance to the Chicago Sanitary and Ship Canal (see Figure 1 and Figure 2). The project site is located behind an existing offshore breakwater that provides partial sheltering during storms (see Figure 2). Existing conditions around Navy Pier can be seen in Figure 2, while Figure 3 shows a rendering of future conditions, including the baseline configuration of the proposed Chicago Gateway Harbor. The proposed development includes one or more new Landing Piers abutting the south edge of Navy Pier, improvements to Dime Pier, new arc-shaped rubble-mound breakwaters extending North and South from Dime Pier, a new rubble-mound breakwater extending north from the Canal entrance, and mooring infrastructure for recreational yachts. Further information on the proposed harbor is available at PBCC (2017).

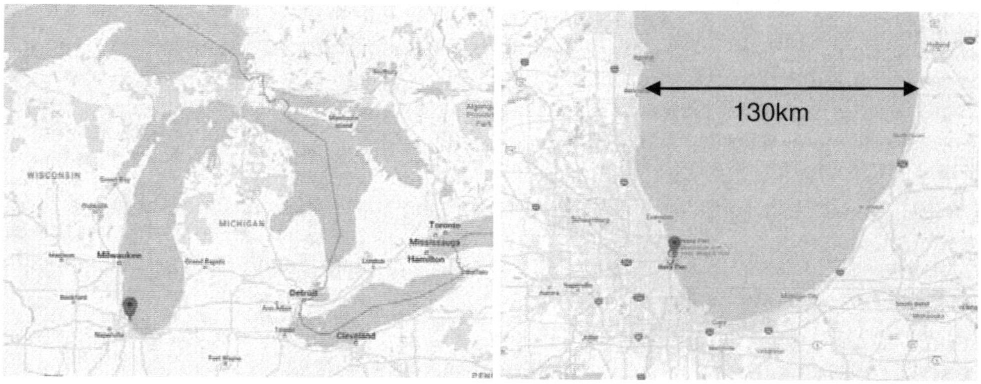

Figure 1. Gateway Harbor is located in central Chicago on the southwest shore of Lake Michigan (images by Google).

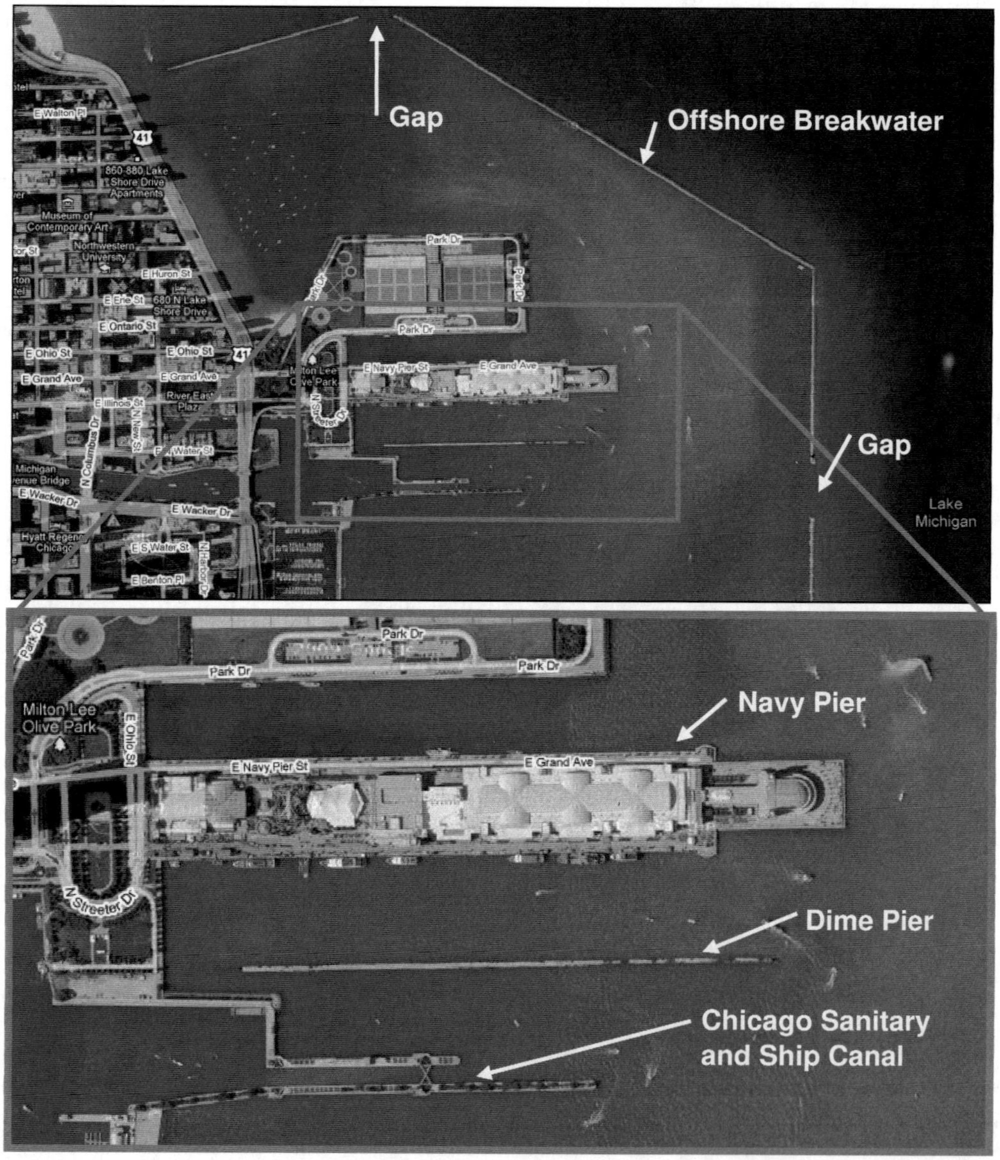

Figure 2. Gateway Harbor is to be located immediately south of Navy Pier in downtown Chicago (images by Google).

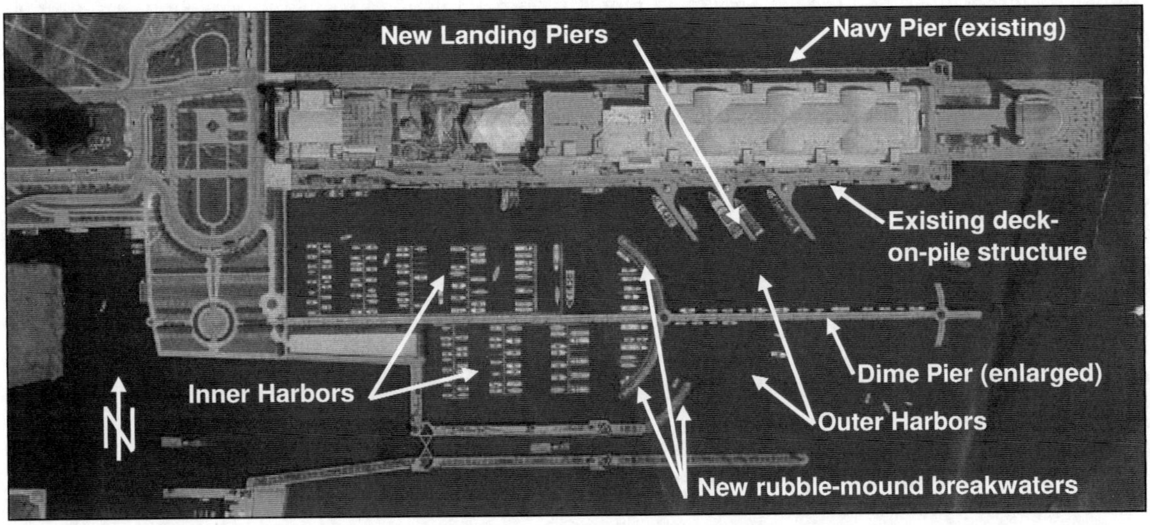

Figure 3. Rendering of the proposed Chicago Gateway Harbor (source: AECOM).

Elevations for this study are referenced to the local Low Water Datum (LWD) which is 1.3 ft (0.4m) above Chicago City Datum (CCD). Lake levels at the site may vary from 0 ft LWD to roughly +6 ft (1.83m) LWD under extreme conditions. Previous analysis by Baird and Associates (1990) suggests that during severe storms, the site may experience significant wave heights up to 5.5 ft (1.68m) approaching from 135° (SE), 112.5° (ESE) and 90° (E) directions, with peak periods of 8 to 10 s. These design conditions result from wave energy overtopping the outer breakwater combined with wave energy penetrating through the gaps in the outer breakwater. The bathymetry throughout most of the harbor area is relatively flat at a depth near -26 ft (-7.9m); becoming gradually deeper heading offshore.

Navy Pier is an important Chicago landmark packed with a wide array of restaurants, shops, entertainment offerings, cultural attractions, and services for tourists. The south side of Navy Pier features a 96 ft (29.3m) wide by 2,429 ft (740.8m) long pile-supported deck structure (relieving platform) that was built in several stages over past years following several different designs. The outer (eastern) end of the deck-on-pile structure features a very dense array of vertical and battered piles. The lakebed under the relieving platform slopes up from an elevation near -25 ft (7.6m) along the south edge to approximately 0 ft next to Navy Pier. The upper part of this fill is protected with armor stone. A portion of the existing deck-on-pile structure can be seen in Figure 4.

Figure 4. Existing deck-on-pile structure along the south side of Navy Pier.

Dime Pier, which is currently in a poor state of repair, will be repaired and enlarged as part of the Gateway Harbor development. The existing pier is approximately 24 ft (7.3m) wide and has a crest elevation near +2 ft (0.61m) LWD. The enlarged pier will be approximately 30 ft (9.1m) wide and will have a crest elevation of +8.0 ft (2.44m) LWD.

The new Landing Piers are proposed as deck-on-pile structures. An option considered in conceptual design was to reduce wave agitation at the new piers by fitting the eastern pier with a vertical wave barrier (baffle wall). However, there was concern that such a wave barrier may reflect waves towards the deck-on-pile structure, thereby causing an increase in the uplift pressures exerted on the underside of the deck under certain conditions. The need for a vertical wave barrier and its effects under design conditions was uncertain and required further study.

The new North and South Arc Breakwaters (attached to Dime Pier) are proposed as a 24 ft (7.32m) wide sheet pile cell with +8.0 ft (2.44m) LWD crest elevation supporting a rubble-mound revetment on the exposed lake side. The new East Breakwater (attached to the north entrance to the Sanitary and Ship Canal) is proposed as a conventional rubble-mound structure with 1:1.5 side slopes, stone armoring, and a 12 ft (3.66m) wide crest at +8.0 ft (2.44m) LWD. Preliminary designs for these structures specified 2.9 and 2.0 ton armor stone.

PHYSICAL MODEL DEVELOPMENT

Study Objectives

The main objectives for the physical model study were as follows:

- Measure wave agitation throughout the outer and inner harbor under the extreme storm conditions particular to the site;
- Measure and observe wave interaction and overtopping at modified existing and proposed new structures within the harbor;
- Measure the wave uplift pressures on the underside of the deck-on-pile structure running along the south side of Navy Pier; and
- Help optimize the initial designs to improve their performance and constructability and reduce overall project costs.

These objectives were fulfilled by designing and constructing a 1/30 scale 3D physical model of Gateway Harbor (including the south side of Navy Pier), outfitting the model with instrumentation for measuring waves and pressures, and by conducting tests to investigate the wave agitation and wave-structure interactions in a range of moderate and severe storm conditions.

Physical Model Design

NRC's 36 m by 30 m Multidirectional Wave Basin (MWB) features a 30 m by 20 m test area and is equipped with a powerful 30 m long 60-segment directional wavemaker. The model was designed according to Froude scaling principles, and scaling relationships derived from Froude scaling laws were used to relate conditions in the model to corresponding conditions in nature. The design philosophy was to adopt as large a model scale as possible to ensure that scale effects resulting from differences in Reynold's number and surface tension would be minimized as much as possible. Freshwater was used in the model to represent the lake water at the project site. The rock materials on the surface of the various rubble-mound structures were sized to have a similar hydraulic stability as in nature.

The physical model was designed so that the outer part of the proposed Gateway Harbor could be modeled and studied in a range of extreme waves approaching from the SE, ESE and E directions. Moreover, the model was designed so that the impacts (if any) of the proposed developments on the uplift pressures exerted on the existing pile-supported deck on the south side of Navy Pier could be assessed. Unfortunately, the entire harbor could not be modeled within the MWB facility at 1/30 scale. However, it was judged better to model most of the new harbor at 1/30 scale, rather than model the entire harbor at 1/50 scale, since the larger scale (1/30) would be technically superior and provide more reliable estimates of both wave disturbance and uplift pressure (see Figure 5).

In general, the bathymetry throughout most of the harbor area is relatively flat at a depth of approximately -26 ft. For sake of economy, the physical model assumed a constant lakebed elevation throughout the harbor area of -26 ft. It should be noted that the outer breakwater and its effects on the incident waves was not included in the physical model domain. Hence, the incident wave conditions for the physical model study were established on the west (shoreward) side of the outer breakwater.

Figure 5. 1/30 scale model of outer Gateway Harbor.

Construction and Outfitting

A 1/30 scale reproduction of the south-eastern part of Navy Pier, including a 1,435 ft (437.7m) long portion of the pile-supported deck, was constructed in the model (see Figure 6). Most of the 96 ft (29.3m) wide deck structure features concrete pile caps on 24 ft (7.32m) centers, each supported by six 3 ft (0.91m) piles. The solid deck is supported by a grid of concrete girders resting on the pile caps. In the model, the girders and pile caps were represented by strips of 19mm and 12mm marine plywood, the piles were representing by plastic tubing, and the deck was represented by sheets of 19mm thick marine plywood or Perspex. The underside of the decking was leveled to an elevation of +6.2 ft (1.89m) LWD to match existing conditions at the site. A sloping revetment was constructed below the deck-on-pile structure to replicate existing conditions. Each of the new 250 ft (76.2m) long Landing Pier structures was modeled in a similar manner, except that the sloping revetment was excluded.

Figure 6. 1/30 scale model of deck-on-pile structure fitted with pressure plate sensors.

Twenty-one pressure plate sensors were installed in the model and used to measure the uplift pressures due to water contact with the underside of the deck-on-pile structures (see Figure 6). The pressure plate design, shown in Figure 7, consists of a circular aluminum sensing plate connected by a stiff rod to a water-proofed high-precision shear beam load cell. The assembly is sealed against water intrusion by means of a thin latex membrane secured with an o-ring. For this study, the pressure plates were installed so that the circular sensing disk was mounted flush with the underside of the deck. Pressures were obtained by dividing the measured load by the effective surface area of the plate (17.5ft^2 or 1.63m^2 at full scale). The pressure plates feature fundamental free vibration frequencies in excess of 500 Hz, making them useful for measuring impulsive wave slamming loads. The load cells were calibrated prior to assembly, and the pressure plates were checked *in situ* by applying a static load. During the study, many of the pressure plate sensors were moved and re-installed several times to record uplift pressures in 46 different locations to suit the objectives of each test series. The pressure plate output was sampled at 500 Hz.

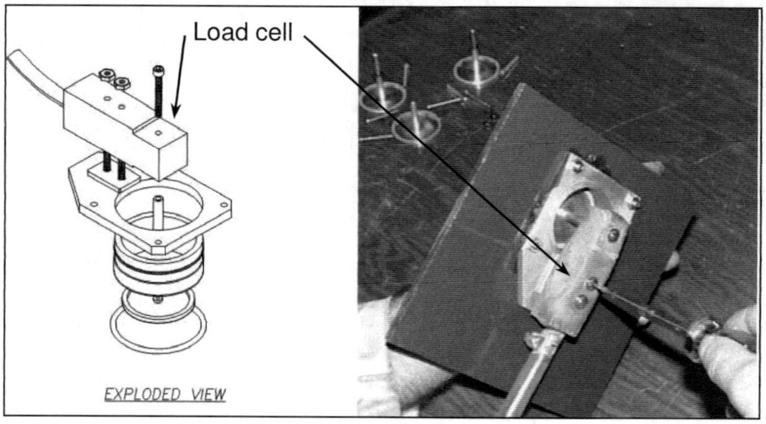

Figure 7. Pressure plate design and assembly.

Masonry blocks were used in the model to simulate the sheet-pile crib structures specified in preliminary designs for the North and South breakwater structures. Five different classes of crushed rock were prepared to match specifications and used to represent the core, filter, and armor layers of the new rubble-mound structures. Templates and precise surveying methods were used to ensure that the model structures conformed to prototype designs. The exposed surface of the armor stone was painted to facilitate visual assessment of damage. Finally, the elevation of the lakebed within the entrance to the southern harbor was replicated in the model by adding a layer of fine gravel on top of the level concrete floor.

Twenty-three capacitance-type wave gauges, which operate by sensing the change in capacitance that occurs as a portion of an insulated wire becomes wetted, were deployed to measure wave conditions at various locations throughout the model domain. The wave gauges were calibrated at regular intervals throughout the study by moving them vertically while the water level remained constant. The measured wave records were analyzed using comprehensive and well-proven GEDAP analysis programs and procedures. GEDAP (Miles, 1990) is a software system developed by NRC to support the synthesis and generation of waves in the laboratory, and for the analysis of waves and wave-related data from physical model studies. Numerous statistics and derived parameters consistent with PIANC-IAHR (1986) definitions were computed from the time histories and spectra recorded at each gauge, and many of these were stored in a spreadsheet database for further analysis, plotting, and manipulation.

Wave Conditions and Water Levels

Thirty-eight unique combinations of water level and incident wave condition were reproduced in the physical model to represent design conditions with return periods ranging from 12 months up to 1,000 years approaching Chicago from the east (90°), east-southeast (112.5°), and southeast (135°)

directions. Water levels ranged from +2.0 ft (0.61m) up to+6.2 ft (1.89m), peak periods ranged from 5.5 s to 9.5 s, and significant wave heights (H_{m0}) outside the outer breakwater ranged from 3.0 ft (0.91m) up to 15.6 ft (4.76m). Wave heights inside (shoreward) the outer breakwater were estimated by AECOM using a combination of numerical simulation and analysis of wave buoy data from both sides of the outer breakwater. AECOM's analysis found that the incident wave conditions inside the outer breakwater tended to vary spatially, with larger wave heights to the south of Dime Pier, and smaller wave heights to the north of Dime Pier. The significant wave height of the incident waves inside (shoreward) the outer breakwater and south of Dime Pier was determined to range from 1.75 ft (0.53m) up to 9.2 ft (2.81m).

All incident seastates were modelled as short-crested irregular waves. The distribution of wave energy with frequency was assumed to follow the well-known JONSWAP spectrum, while the distribution of wave energy with direction was assumed to follow a cosine-power spreading function. The directional spreading functions for this study had a standard deviation near 15°. Target wave time histories were synthesized from the target directional spectrum using the single summation model paired with the random phase method as described by Cornett et al. (1993). For this study, the duration of each wave train was set equal to one full scale hour (or 657 s at model scale).

For this study, in order to better simulate the spatially varying wave conditions inside the outer breakwater near the end of Dime Pier, the wavemaker was notionally divided into three zones, and different wave conditions were synthesized and generated in each zone. The northern zone included wavemaker segments 1 to 20 and was used to generate the wave conditions specified on the north side of Dime Pier. The southern zone extended from wavemaker segment 27 to 60 and this part of the machine was used to generate the wave conditions specified on the south side of Dime Pier. A smooth linear transition between the two different wave fields was specified over the third zone, between wavemaker segments 21 and 26. All of the incident wave conditions were generated, measured and verified in the physical model before constructing any of the new structures. The wavemaker command signals were tuned so that the incident wave conditions (T_p, H_{mo} and mean direction) measured on the south and north sides of the outer end of Dime Pier were in good agreement with specifications.

Standard GEDAP time-domain and frequency-domain analysis algorithms were applied to analyze in considerable detail the wave conditions measured in the model. The directional properties of the incident waves were estimated using the Surface Slopes Maximum Entropy Method (SSMEM) described by Cornett et al. (2005), based on wave data measured at a compact 4-gauge array.

PHYSICAL MODEL OUTPUT

Test Program

In additional to modelling existing conditions, over twenty alternative layouts for Gateway Harbor were modelled and assessed in this study. Layout 1 simulated existing conditions (without the new harbor structures), while layouts 2 through 23 simulated alternative realizations of the Gateway Harbor development. Some of the layouts were quite similar to each other, while others featured significant differences. For the most part, these alternative layouts were explored while attempting to find an optimal layout that was economically viable, was able to attenuate the peak uplift pressures recorded on the underside of the existing deck-on-pile structure along the south side of Navy Pier, and was able to provide adequate protection for the inner harbor. For many of the layouts, vertical walls of various lengths were added to the southern edge of the deck-on-pile structure in order to block waves from penetrating under the platform. Harbor layouts 7 through 18 included a new rubble-mound breakwater structure attached to Navy Pier and located just beyond the eastern end of the deck-on-pile structure (see Figure 8a), which was not included in the baseline harbor layout. For layouts 19, 20, and 21, the new breakwater attached to Navy Pier was removed and a vertical wall was added to close off the east (outer) end of the deck-on-pile structure. The eastern Landing Pier was fitted with a solid vertical wall for harbor layouts 2, 4, 5, and 21. The western Landing Pier was fitted with a solid vertical wall for harbor layouts 3 and 14 (see Figure 8b).

a) 
b) 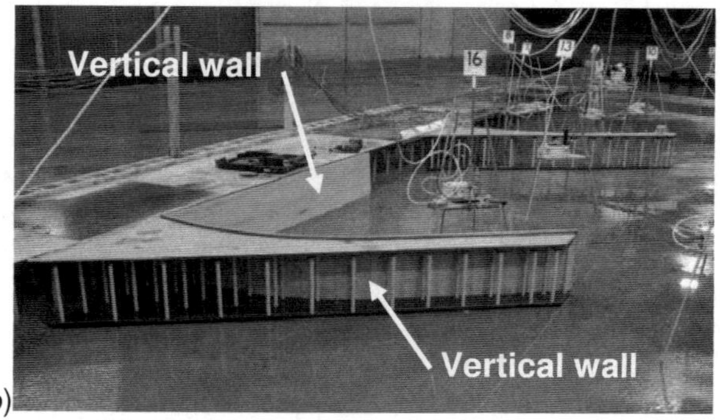

Figure 8. a) New spur breakwater attached to Navy Pier for harbor layout 8;
b) vertical walls added to Landing Pier and Navy Pier deck-on-pile structure for harbor layout 3.

Wave Disturbance

Wave conditions at multiple locations in the inner and outer harbors were measured for each of the harbor layouts described in the preceding section. In most cases, the performance of each layout was assessed for a few of the most critical combinations of water level and incident wave condition. In many cases, the changes in harbor layout considered in these tests had little influence on the wave conditions throughout many parts of the harbor. However, significant changes were sometimes observed at specific locations. Figure 9a shows an example of the wave condition measured at 20 locations in the physical model for a single test condition and nine different harbor (port) layouts. The measurement locations are shown in Figure 9b. Stations WP20-23 and WP01 are located on the south and north sides of the head of Dime Pier, WP15 is located within the south Inner Harbor, and WP19 is located within the north Inner Harbor. For this wave condition, wave heights in the outer harbour near the head of Dime Pier ranged from 6 ft (1.83m) north of Dime Pier) to ~7.4 ft (2.26m) south of Dime Pier. Wave heights in the south Inner Harbor (WP15) were steady at 2 ft (0.61m) for all layouts, while wave heights in the North Inner Harbor (WP 19) varied from 1.5 ft (0.46m) up to 4.5 ft (1.37m), depending on the harbor layout. The greatest variability in wave height occurred at locations WP07 and WP08, both located near the edge of the pile-supported deck structure and adjacent to the eastern Landing Pier, and at location WP03, near the southeast corner of the deck-on-pile structure. Wave heights in these three locations were larger for layouts where a vertical wall was added to the edge of the existing deck-on-pile structure in these locations. Results from three repetitions of a single test condition indicated that the wave conditions observed in the model were highly repeatable. The average standard deviation in significant wave height at each wave gauge was 0.10 ft (31mm), with a minimum standard deviation of 0.02 ft (6mm) and a maximum of 0.21 ft (64mm) full scale.

Compared to existing conditions, the baseline Gateway Harbor layout was responsible for significant reductions in wave agitation within the Inner Harbors over a wide range of moderate and extreme conditions. The addition of a vertical wall to one of the Landing Piers (see Figure 8b) caused the wave conditions in the vicinity of the Landing Pier to amplify and become more erratic; however, the vertical wall also served to reduce the amount wave energy able to penetrate into the inner portion of the North Harbor. This wave height reduction within the inner harbor was greatest when the vertical wall was added to the western Landing Pier, but was also significant when the eastern Landing Pier was fitted with a vertical wall.

The various forms of spur breakwater attached to Navy Pier studied in port layouts 7 – 18 (see Figure 8a) also provided notable reductions in overall wave agitation within the harbor for waves approaching from the east, but was less effective at attenuating waves approaching from more southerly directions.

The test results clearly showed that the wave agitation in the south Inner Harbor remained unchanged after not only shortening the South Arc Breakwater (harbor layout 6) but also after removing a significant portion of the armor stone on the South Arc Breakwater (harbor layout 13). These alterations represent a significant cost savings without any noticeable penalty on wave agitation.

Although quantitative measurements of wave overtopping were not collected, it was clear from qualitative observations that the pile-supported deck running along the south side of Navy Pier was regularly awash with water during the most severe test conditions. The overtopping was most frequent and most severe at the eastern end of the deck-on-pile structure. Significant sheet flows were observed across the deck throughout the duration of the worst storm events. Harbor layouts that included a new spur rubble-mound breakwater attached to Navy Pier were able to reduce this overtopping appreciably; however, it was not totally abated.

Significant levels of wave overtopping were also observed on the easternmost Landing Pier and near the junction between the Landing Pier and the existing deck-on-pile structure. The frequency and intensity of this overtopping was increased when the eastern side of the Landing Pier was outfitted with a vertical wall.

Further design work should bear in mind the potential risk to pedestrians and infrastructure due to overtopping flows during large storm events. It is noted that the wave agitation and wave overtopping may also be exacerbated by boat wakes, which were not simulated in these studies.

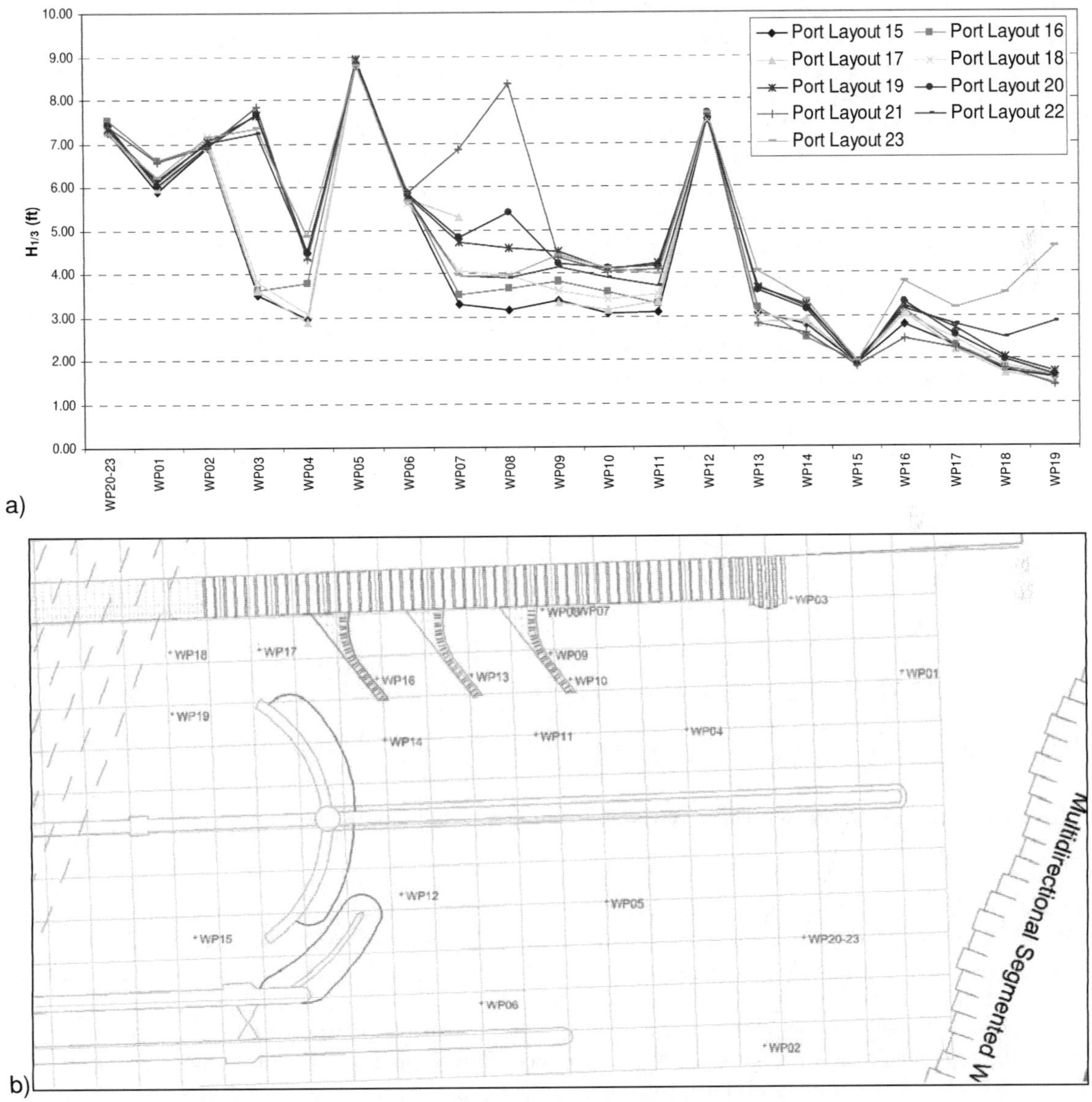

Figure 9. a) Influence of harbor layout on significant wave heights for seastate SE6; b) map of wave gauge locations.

Wave Uplift Pressure

A main focus of the testing was to determine the influence of the Gateway Harbor development on the wave uplift pressures on the pile-supported deck running along the south side of Navy Pier. Once this was established for the baseline harbor layout, the focus shifted to investigating alternative harbor layouts that would alleviate the uplift pressures and reduce wave agitation and wave overtopping as much as possible without increasing project costs substantially. Pressure readings were recorded at a rate of 500 samples per second (model scale) during every test conducted with harbor layouts 1 through 23. Uplift pressures were recorded in 46 locations using 21 pressure plate sensors. The pressure time histories were analyzed to determine various statistical quantities such as the maximum pressure, the average value of the five largest independent pressure pulses, and the average value of the ten largest independent pressure pulses. The latter two statistics provide less extreme, yet more statistically stable and reliable measures of the uplift pressure at each location. The average value of the five largest independent maximums can be thought of as the expected maximum over a 12 minute (prototype) duration. Similarly, the average value of the ten largest maximums is equivalent to the expected maximum over a 6 minute (prototype) duration.

Two representative examples of the pressure pulses recorded on the underside of the deck-on-pile structure are shown in Figure 10. Each figure shows the time history of the maximum pressure event (in black) together with the pressure fluctuations at neighboring sensors located nearby (colored lines). It should be noted that these pressure pulses are expressed in kilopascals (kPa) at full scale. It was evident that the largest uplift pressures were highly dynamic and impulsive and had short durations. Also, the pressures at neighboring sensors generally did not peak at the same time. In many cases the peak pressures at nearby locations were also much smaller. This implies that for any instant in time, the area of deck subject to high uplift pressures is localized and small.

Since the largest pressures were localized and acted over a very short interval, their impact may well be absorbed and dissipated within the deck-on-pile structure before any damage can occur. A dynamic analysis of the structure is recommended to properly interpret the impact and consequence of the impulsive pressures observed in this study.

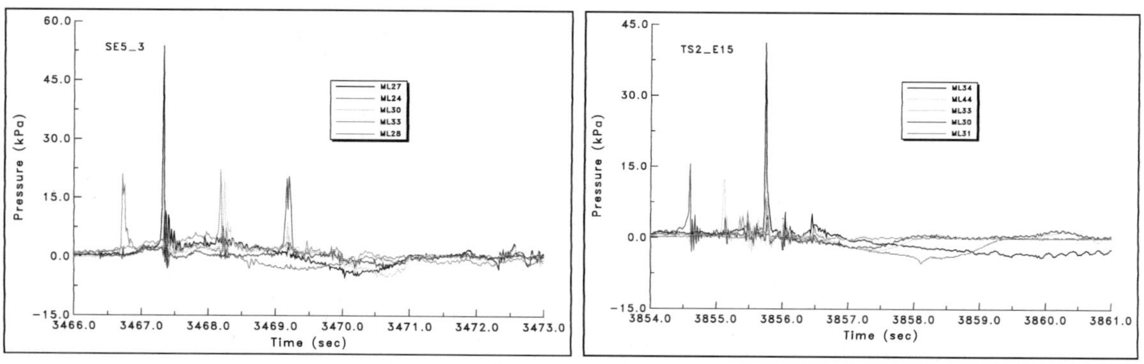

Figure 10. Examples of uplift pressure pulses recorded during the Gateway Harbor study.

AECOM undertook a separate investigation to determine the load capacity of the Navy Pier deck-on-pile structure against uplift, which was estimated to be approximately 400 psf (19.2kPa). The easternmost part of the platform, constructed after 1991, was estimated to have a loading capacity against uplift of approximately 700 psf (33.5kPa). The pressures recorded in tests with the model configured to simulate existing conditions suggested that the existing deck-on-pile structure may have previously, or could in future, experience peak wave-induced uplift pressures nearing or exceeding the available loading capacity under severe storm conditions. The largest pressures were observed near the southern (outer) edge of the platform.

Results from tests with the baseline Gateway Harbor layout indicated that, for many test conditions, the peak uplift pressures on the deck-on-pile structure near the east Landing Pier were greater than for existing conditions. This pressure increase can be attributed to an increase in wave energy at the deck-on-pile structure caused by waves reflecting from the solid vertical wall installed on the curved east side of the eastern Landing Pier, as well as reflections from the North Arc Breakwater. Since the pressure sensors were concentrated in one area for these tests, it was difficult to discern the spatial extent of the pressure increase. For the next series of tests, the vertical skirt wall was removed from the eastern Landing Pier and installed on the western Landing Pier. At the same time, a 240 ft (73.2m) long skirt wall was installed along the southern edge of the deck-on-pile structure, near the junction with the western Landing Pier. The peak uplift pressures for this third harbor layout were smaller than for layout 2, but the wave conditions in the outer harbor were larger and more chaotic due to the wave energy reflecting from the vertical walls. In subsequent testing, several different arrangements of vertical skirt walls were modelled and assessed, while the 21 pressure plate sensors were moved to different locations to obtain information on uplift pressures on other regions of the deck-on-pile structure. A new rubble-mound spur breakwater extending south from the east end of Navy Pier was modelled in harbor layouts 7 through 18 (see Figure 8a), and this new structure proved effective in attenuating uplift pressures without exacerbating wave conditions in the outer and inner harbors for waves approaching from the east, particularly when the breakwater crest elevation was increased from +6 ft to +8 ft (1.83m to 2.44m). However, the new spur breakwater was less effective at attenuating waves approaching from the SE and ESE directions. Dime Pier was extended by 115 ft (35.1m) for harbor layouts 10 and 11 and this modification helped attenuate wave energy approaching from the SE and ESE directions.

Some of the more effective harbor layouts were able to reduce uplift pressures below the levels observed for existing conditions. Figure 11 shows the changes in peak uplift pressure measured at eleven locations near the east end of the deck-on-pile structure in tests with harbor (port) layouts 4 through 14 and identical storm waves approaching from the east. This figure illustrates how optimizing the new harbor layout was able to reduce the largest uplift pressures to less than one-third of their earlier values. The finding that excessive uplift pressures could be reduced to acceptable levels by making relatively small and inexpensive changes in harbor layout was an important factor contributing to the viability of the Gateway Harbor development.

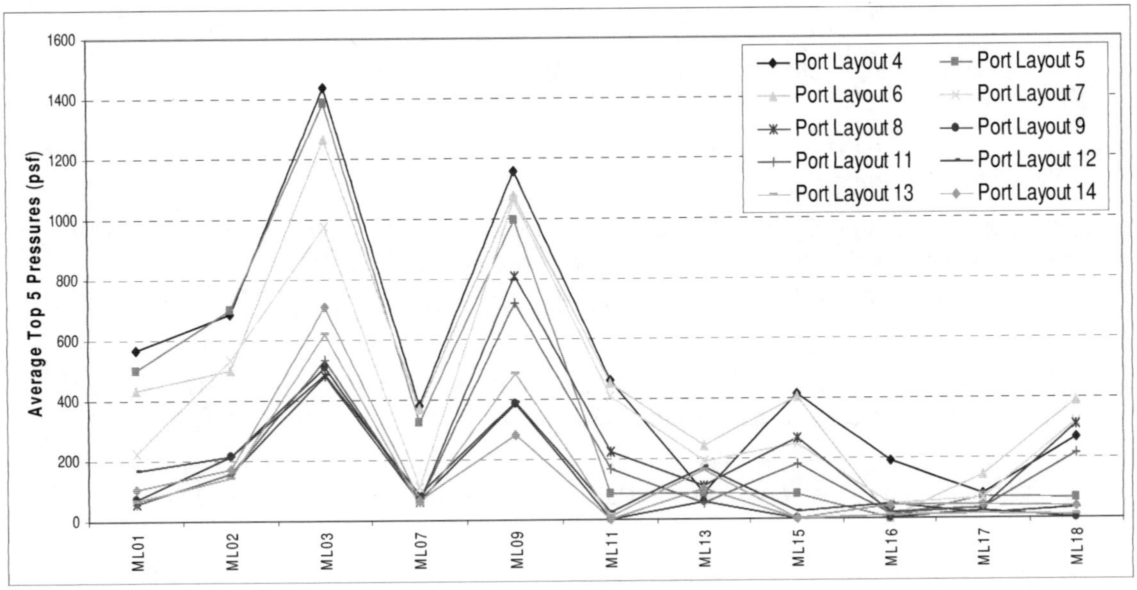

Figure 11. Influence of harbor (port) layout on peak uplift pressures at the east end of Navy Pier for storm waves from east.

CONCLUSIONS

The important role that physical modeling can play in supporting the efficient and optimized design of ports, harbors, and marinas has been discussed in this paper with reference to a specific project example known as Gateway Harbor, Chicago. In this project, testing in a 1/30 scale 3D physical model played a key role in assessing wave loads, wave disturbance, and wave overtopping for the baseline layout, and in developing, assessing, and validating important design optimizations that could alleviate uplift pressures and reduce wave agitation and wave overtopping as much as possible without increasing project costs substantially. Over twenty alternative harbor layouts were modeled and assessed in the physical model study in a highly-efficient and cost-effective manner.

Some of the more effective harbor layouts were able to reduce uplift pressures, agitation levels, and overtopping considerably compared to the baseline layout. The improved performance resulting from these optimizations was important for the overall viability of the Gateway Harbour project, and would not have been possible without the 3D physical model study. The results of the study are being used by AECOM to develop a detailed final design for Chicago Gateway Harbor which reflects an optimized balance of cost, performance and risk.

REFERENCES

Cornett, A., Anglin, D., Elliot, T., 2013. Wave-in-Deck Loads for an Intricate Pile-Supported Pier and Variation with Deck Clearance. *Proc. 32nd Int. Conf. on Ocean, Offshore and Arctic Eng.* (OMAE-2013), Nantes, France.

Cornett, A., M.D. Miles, & D. Pelletier. 2005. Measurement and Analysis of Multidirectional Waves Using Free Surface Slopes. *Proc. 5th Int. Symposium on Wave Measurement and Analysis* (WAVES 2005), Madrid, Spain.

Cornett, A., Miles, M.D. & Nwogu, O.G. 1993. Physical Modelling of Multidirectional Waves. *Proc. 2nd Int. Symposium on Wave Measurement and Analysis* (WAVES'93), New Orleans, USA.

Cuomo, G., Tirindelli, M., Allsop, W., 2007. Wave-in-deck loads on exposed jetties. *J. Coastal Eng.* 54 657-679.

Cuomo, G., Shimosako, K., Takahashi, S., 2009. Wave-in-deck loads on coastal bridges and the role of air. *J. Coastal Eng.* 56 (2009) 793-809.

Meng, Y., Chen, G., Yan, S. 2010. Wave interaction with deck of jetty on a slope. *Proc. 2010 Int. Conf. on Coastal Eng. (ICCE 2010)*, Beijing, China.

Miles, M.D., 1990. The GEDAP™ Data Analysis Software Package. NRC Technical Report TR-HY-030, Ottawa, Canada.

Murali, K., Sundar, V., Setti, K., 2009. Wave-induced pressures and forces on deck slabs near the free surface. *J. Waterway, Port, Coastal and Ocean Eng.* 135:269-277.

PIANC-IAHR (1986) List of Sea State Parameters. *Supplement to Bulletin 52*, PIANC, Brussels, Belgium.

Public Building Commission of Chicago (PBCC), 'Gateway Harbor', 2017. [Online]. Available: http://www.pbcchicago.com/projects/gateway-harbor. [Accessed 20 March 2018].

Tirindelli, M., Cuomo, G., Allsop, W., Lamberti, A., 2003. Wave-in-deck forces on jetties and related structures. *Proc. 13th Int. Offshore and Polar Eng. Conf.* (ISOPE 2003), pp 562-569.

GEOTEXTILE TUBE AND GABION ARMOURED SEAWALL FOR COASTAL PROTECTION AN ALTERNATIVE

by

S Sherlin Prem Nishold[1], Ranganathan Sundaravadivelu [2*], Nilanjan Saha[3]

ABSTRACT

The present study deals with a site-specific innovative solution executed in the northeast coastline of Odisha in India. The retarded embankment which had been maintained yearly by traditional means of 'bullah piling' and sandbags, proved ineffective and got washed away for a stretch of 350 meters in 2011. About the site condition, it is required to design an efficient coastal protection system prevailing to a low soil bearing capacity and continuously exposed to tides and waves. The erosion of existing embankment at Pentha (Odisha) has necessitated the construction of a retarded embankment. Conventional hard engineered materials for coastal protection are more expensive since they are not readily available near to the site. Moreover, they have not been found suitable for prevailing in in-situ marine environment and soil condition. Geosynthetics are innovative solutions for coastal erosion and protection are cheap, quickly installable when compared to other materials and methods. Therefore, a geotextile tube seawall was designed and built for a length of 505 m as soft coastal protection structure. A scaled model (1:10) study of geotextile tube configurations with and without gabion box structure is examined for the better understanding of hydrodynamic characteristics for such configurations. The scaled model in the mentioned configuration was constructed using woven geotextile fabric as geo tubes. The gabion box was made up of eco-friendly polypropylene tar-coated rope and consists of small rubble stones which increase the porosity when compared to the conventional monolithic rubble mound. In such a configuration, multi-tiered geotextile tube seawall was constructed with four layers of 10 hydraulically filled geotextile tube as the core, while stone filled polypropylene tar coated rope gabion boxes acted as armour layer for the structure. This scaled model examined for emerged water conditions of 0.5 m design water depth for different wave heights and different wave periods. The geotextile tube with gabion showed good wave energy dissipation characteristics. Furthermore, reflection characteristics of this model were also quantified. After that, the design was implemented and constructed as a pilot project on Pentha coast. This case study establishes geotextile tube seawall as an alternative to the conventional method of coastal protection.

Keywords: Seawall, Geotextile tube, Coastal Protection, Gabion, Embankment.

1 INTRODUCTION

The coastal state of Odisha is almost protected with saline embankment for a length of 475 km along the shoreline, which constructed with locally sourced soil. A particular stretch of saline embankment has been observed to regularly eroded during the storm surge, tides, waves, and flood. Pentha (20°32.5'N 86°47.5'E) is a coastal village in Kendrapara District of Odisha State at about a distance of 8.6 km from Rajnagar Town, in India. The damage to the saline embankment was posing a significant threat to the lives and livelihood of the coastal communities. In addition to this, As per the past 25 years, metrological data pertain to the coastline was also affected by two cyclones, viz. Phailin (2013) and Hud Hud (2014). Therefore, a retarded embankment built which is also likely to erode if not protected. The Government of India intends to construct a suitable geotextile tube embankment on the seaside of the retarded embankment. Hence a new geotextile tube embankment was proposed which lies between the two points of (20°32'23.10" N – 86°47.18.01" E) and (20°32'23.10" N – 86°47'18.01" E) for 505 m length (Figure 1). The site was continuously affected by cyclones and storm surge, associated with a low pressure weather system, whereas the tidal ingression is around 500 meter into the land since 1999, that causes the water to pile up higher than the ordinary sea level and tends to increase the wave height which is a predominant reason for erosion of beach berms and dunes, since storm surge waves are non-breaking waves. The region was connected with Hexa Rivers Brahmani, Baitarani, Chinchiri, Pathsala, Maipura, Kharasrota, Barunei and Dhamara. The coastal tracts with those rivers are interconnected with fault lineament. The general topography is irregular with many drain cuts, rivers, lakes, ponds, swamps, estuaries and lagoons.

[1] Ph.D Research Scholar, Department of Ocean Engineering, Indian Institute of Technology Madras, India 600 036.
[2] Professor, Department of Ocean Engineering, Indian Institute of Technology Madras, India 600 036, rsun@iitm.ac.in
[3] Associate Professor, Department of Ocean Engineering, Indian Institute of Technology Madras, India 600 036.

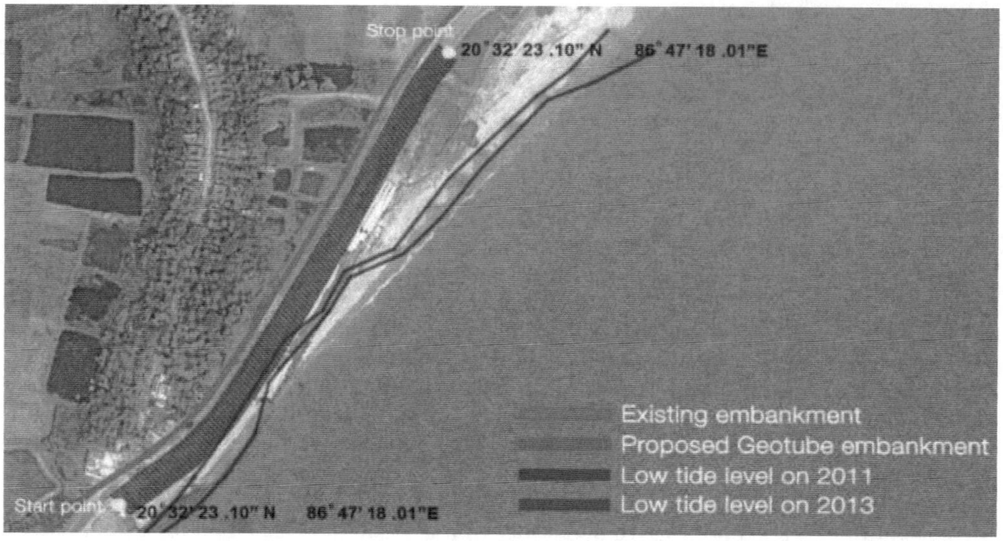

Fig. 1 Google image of geotextile seawall of 50m scale (accessed on July 2017)

1.1 Coastal erosion

High tide level at the site is about 4 m w.r.t MSL and storm surge is 1 m. Therefore, vast quantities of tidal reach pass into rivers for more than 20 km distance from the river mouth. The site lies between two rivers which discharge water into the sea and the circulation of currents between these two river clusters lead to erosion. The bathymetry is perfectly parallel to the shoreline, and the beach slope is 1:60, resulting in the formation of regular waves at equal intervals. Since beach slope is gentle wave breaks on the longshore bar, and due to higher wave celerity, it plunges over foreshore up to berm. Which results in movement of sediments from onshore and transported back to foreshore during backwashing, In addition to cross-shore transport (littoral drift). Further, the site is continuously affected by cyclones, storm surges, associated with a low-pressure weather system. A storm surge causes the water to pile up higher than the average sea level and tends to increase the wave height. It is the predominant reason for the erosion of beach berms and dunes. Since storm surge waves are of high intensity and breaks after longshore bar, the gradient in transport rate increases in the direction of net transport. The conventional materials usually used for protection against coastal erosion are rubble mounds and artificial armour units. However, these materials are costly and time-consuming to install, apart from not being readily available in large enough quantities. On the other hand, geosynthetics (geotextile-tube and gabion embankments) can serve as cost-effective soft engineering solutions for coastal protection. Moreover, when compared with conventional rubble mounds (core stone of density 2.65 t/m^3 with 20% void ratio) the load intensity due to geotextile tube (filled with sand of density 2 t/m^3 with 30% void ratio) is significantly lower. Additionally, geotextile tube acts as a single monolithic unit of high stability whereas core stones will be heterogeneous. Gabion box filled with small rocks which is more porous and hence, highly dissipative to wave energy compared to armour stones of equal weight. Thus, geotextile tube with gabion box protection is an excellent solution for coastal protection applications. The increase in wave reflection on coastal structures can lead to poor performance under the rough weather, which results in the increased possibility of scour and failure of coastal structures. One way of solving such problem is by deploy embankments with high energy dissipation characteristics, using unique geometries of geotextile tube and gabion embankments. In this study, an attempt is made to design and to understand the hydrodynamic characteristics of geotextile tube along with gabion boxes as armour layer.

1.2 Process of erosion

Coast near the Pentha is a village subjected to severe erosion for the past 25 years. Initially, the sea was 500 m away from the existing saline embankment. Since this original saline embankment eroded, a retarded embankment has been constructed 60 m behind. The shore was at 50 m from the retarded embankment on 21st Nov 2009 and on 23rd Oct 2011 and coastline was at 33 m from retarded embankment eroded at a rate of 8.5 m/annum. Hence the erosion rate is about 8.5 m per annum. Storm waves from 2009 encroached around 300 m stretch of the retarded embankment. Since the retarded embankment proven ineffective a new standalone geotextile tube embankment has designed with 30 m base width and a height of 7.4 m, aligned about 5 m to 10 m away from the retarded embankment for a length of 700 m during 2011; The standalone geotextile tube cross-section detailed in Figure 2. However, due to the subsequent erosion of coast, the base was integrated, and the width of geotextile tube embankment had altered to 24 m from 30 m.

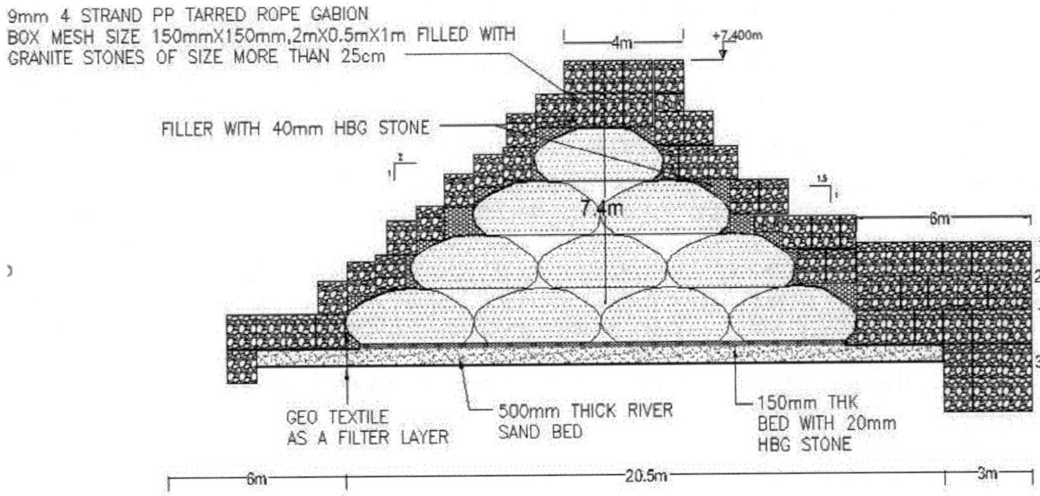

Fig. 2 Typical cross-sections of geotextile tube embankment (All dimensions are in m)

1.3 Site details

Soil samples obtained from the site was air dried, pulverised and sieved. The soil properties obtained through various laboratory tests. Soil samples collected at different locations reveal that the clay proportions in the soil content are about 87% and remaining silt and fine particles of quartz ranging 1 μm to 0.6 μm. The black coloured clayey soil in this region consists of pods and pockets; these soil samples can induce high hydrostatic pressure in the fluid within pore thus causing impounding of groundwater. The obtained soil samples are highly plastic with a liquid limit of 82%, plastic limit of 36% and the with a plasticity index of 46%, such soil is semi-impermeable, which is capable of absorbing large amounts of water due to the structural, absorption and capillary effects. The soil was classified as CH (Clay of highly plastic) with 87% of clay and 13% of silt based on unified soil classification system.

2 Background

Hydrostatic pressure occurs due to groundwater seepage and development of pore water pressure mobilises sufficient stress in the soil. The foundation soil over the site is highly plastic, and the soil characteristics are poor. Hence this low bearing and undrained material behaviour can offer less resistance to the structural loads when exposed to dynamic wave loading. Apart from foundation soil, wave characteristics must adequately assess through model study, which needed for understanding structural stability well in advance. A brief literature study of geotextile tube seawall with and without gabion protection as a coastal structure conducted along with the practical experience gained on the construction grounds.

2.1 Geotextile

Geotextile is synthetic material available in the woven and non-woven form, with various material compositions such as polyester, polyethene, and polypropylene. These materials are eco-friendly and non-reactive to the marine environment. Geotextile fabricated into various elements such as geotextile bags, geotextile mattress, geotextile tubes and geotextile containers. Each one of it will differ on its geometry used for different applications depending upon the loads, the strength of the fabric used, filling method, filling ratio, stability, and durability. In this regard, (Heerten and Wittmann 1985) discussed the physical dimensions of geotextile, the gradation of fill material and filter criteria based on the geotechnical application related to river and canal application. A complete physical and chemical laboratory test on geotextiles to assess the permeability and soil retention is provided by (Luettich et al. 1992).

2.2 Geotextile tubes

Geotextile tubes are made up of synthetic fibres which are sustainable, permeable textile fibres those can contain, filter, and reinforce soil. The integrity of the geotextile structure depends on the type of infill material and type of geosynthetics used. The permeability of the infill material and Apparent Opening Size (AOS) of geotextile has significant impacts on water outflow and rate of formation of the filter cake. Consequently, the strength of the soil infill in geotextile tubes with high moisture content will not be sufficient to support geotextile tube stacking (Shin and Oh 2007). Leshchinsky *et al.* 1996 have developed an analytical solution based on a computer program (GeoCops), to predict the design parameters such as pumping pressure, circumferential tensile force, and unit weight of the fill material along with the tube height. Various studies on the stability of stacked geotextile tubes under wave actions can be found in the works of Van Steeg et al. (2011). Experimental studies of geotextile sand-filled containers for dune erosion have been carried out by (Das Neves et al. 2009, Bezuijen et al. 2004, Pilarczyk 2008, Cho 2009). Kim *et al.* (2013) performed Finite Element Analyses (FEAs) on ground modification techniques for improved stability of geotextile tube–reinforced reclamation embankments subject to scouring. However, there are few studies on the stability of stacked geotextile tubes subjected to hydrodynamic characteristics and scouring.

2.3 Polypropylene rope gabion boxes

Gabion boxes filled with a smaller range of stones are more porous and therefore capable of dissipating sizeable kinematic wave forces. Stacking of gabion boxes with each other in various interlocking patterns is equivalent to installing the armour units for the conventional constructions (Motyka and Welsby 1987, D'Angremond et al. 1992, Takahashi 1997). U.S Army Crops, (1986) describes the use of gabions in the coastal environment subjected to wave forces and saltwater corrosion. The design of stepped gabion method of construction methods of spillways including gabion suitability and the hydraulic performances were investigated experimentally regarding the flow patterns, air-water flow properties, and energy dissipation (Wuthrich and Chanson 2014). For the present study, flexible tar coated polypropylene gabion box is used to protect stacked geotextile tube core, and gabion shield will act as armour. These gabion boxes will dissipate the wave energy because of its porous nature. It helps in scour protection and integrity of the geotextile tube core. The gabion was placed layer by layer in the form of English bond brickwork technique and correctly laced together horizontally and vertically using polypropylene tarred rope after the stacking of gabion box in position. All the gabion boxes have tied each other manually to the adjoining boxes on all sides. This arrangement will protect the gabion boxes from movement in case of large wave forces. If any differential settlement of the soil occurs, the geotextile tube will adjust with soil bed profile because of the flexibility and porous nature of geotextile tube fabric. The geotextile tube embankments also protect the inland area from erosion and stormwater inundation and provide proper coastal protection from severe in-situ erosion. Further, it facilitated by a scour apron that has been designed to protect stacked geotextile tube. Heavy-duty plastic mesh type of gabion boxes are not used in coastal protection, hence its effectiveness to tested.

2.4 Factors influencing the stability of geotextile tube and gabion boxes

There are two major factors for the failures of geotextile tube structure: the hydrodynamic factors (such as inertia and drag) and geotechnical factors. The wave-induced lateral forces must counterbalance in addition to horizontal and vertical loads by the geotechnical characteristics of soil bed profile. Inadequate handling of these loads can lead to various types of failures of the geotextile tube embankment system. The factors influencing the stability of geotextile tubes are detailed in following sub-sections.

2.4.1 Hydrodynamic failure mechanism

Hydrodynamic loads can alter the shape and geometry-related characteristics of the geotextile tube configuration, locally as well as globally. This effect set up various mechanisms of failures as reported by (Jackson *et al.* 2006) and (Lawson 2008). These studies show that sand loss or sand migration is due to the aggressive action of waves and currents which passes through the geotextile pores. This cause the failure of the geotextile tube containment system. These losses can detect a loss of sand fineness within the geotextile tube cross-section. The rate of sediment loss from the geotextile tube structure will initially influence the structure geometry, and in due time it will fail. To prevent the sand loss, the particle size of fill soil should be higher than the geotextile aperture size. Another reason for sediment loss (although not directly related to hydrodynamical loads) can be due to the damage of geotextile such as vandalism, bursting, and puncturing. The different types of hydrodynamic failure mechanism of geotextile tube due to sand loss are as follows.

2.4.2 Geotechnical failure mechanism

Geotechnical failures refer to the failure of the base or sub-base layer underneath the geotextile tube. Hence, such failures depended upon engineering characteristics and physical composition of soil which will vary concerning location, environment and influence of load acting upon them. Usually, engineering properties which modified during soil deformation are the shear strength, stiffness, and permeability. Coastal structures exposed to wind, waves and currents; hence these environment characteristics also influence the foundation soil properties and its stability. Such scenarios explained in detail in following sections.

2.5 Need for the geotextile tube embankment

Geotextile tube made of woven geotextile sheets which are flexible and perforated, hence allows water to exit. Thus the development of pore water pressure will be avoided. If any differential settlement occurs, the geotextile tube will adjust with soil bed profile because of the flexibility and porous nature of geotextile tube. These tubes will act as a solid core in the embankment and will serve as an impervious medium. However, these geotextile tube elements continuously subjected to various static and dynamic forces such as gravity, surcharge, and wave loads. In addition to the lateral forces, they support overburden pressure. A combination of these applied forces and loading may contribute to potential problems. Therefore, to counterbalance these forces gabion boxes are used to dissipate the large kinetic wave forces. Another significant advantage of the gabion boxes is that they shield the geotextile fabric from solar Ultra-Violet (UV) radiation and hence increase their durability.

3. Experimental Investigation

To understand the hydrodynamic behaviour of geotextile tube embankment, a series of experiments have been performed for two different models. Geotextile tube standalone Structure and a second model of Geotextile tube with gabion structure were installed.

3.1 Test Facility

The experiments conducted in a wave flume at the Department of Ocean Engineering, Indian Institute of Technology Madras, India. The flume is 72.5 m long, 2 m wide and 2 m deep. A hydraulic piston wavemaker is installed at one end of the flume and has been used to generate waves with predefined characteristics for these set of experiments. A personal computer, connected to the servo actuator was used to input the time history of the signal to the wave maker as well as for the data acquisition of the signals from wave gauges through an amplifier. An artificial beach consisting of a combination of a parabolic perforated steel sheet and a rubble mound is provided at the other end of the flume to absorb the generated waves efficiently.

3.2 Details of Prototype and Scaled Model

Sand filled geo-tubes and geo-bags will be an alternative source for coastal erosion and scour protection in the case where the conventional rubble mound and another kind of artificial concrete armour units cannot use as a protective measure in various circumstances, such as low bearing capacity, severe erosion, flooding. Erosion of shoreline is a predominant phenomenon which takes place because of movement of sand mass by wave action, tidal currents, and wave-induced currents. The conventional material of coastal protection is not only expensive and time-consuming, but this material may not be readily available. Geosynthetics are innovative solutions for coastal erosion and protection which are cheap and quickly installable when compared to other conventional materials and methods. This paper discusses different approaches for construction of geo-tube embankment as a coastal protection structure using geosynthetics. Detailed scaled model studies of scale 1:10 of the geo-tube embankment with ten geo-tube of four-layer has studied without and with gabion protection. The model base width is 1.2 m and with a crest width of 0.25 m. Various tests were performed for different water depth such as maximum water depth of 0.5m and wave height of 0.1 m, 0.2 m and 0.25 m for wave period range of 0.6 sec to 2.5 sec. The prototype geo-tube embankment parameters had been scaled to model using Froude scaling. Using a chosen scaling ratio of 1:10 and model dimensions had arrived, and details had furnished in Table 1.

Type of Structure	Prototype	Model (1:10)
Geo-tube Circumference	9 m	0.9 m
Geo-tube Diameter	3 m	0.3 m
Gabion Box Dimension	2 m x1 m x 1 m	0.2 m x 0.1 m x 0.1 m
High Tide Level (HTL)	4 m	0.4 m
Strom Surge (SS)	1 m	0.1 m
Maximum Height of Water Depth (D max) = HTL + SS	5 m	0.5 m

Table 1. Details of Prototype and Scaled Model

3.3 Model Setup and Test Condition

The positions of the wave gauges and the erected model in the wave flume shown in Fig 3, and 4 show the top view and side view, respectively, of the proposed model in a wave flume. The length of an individual geotextile tube structure and geotextile tube with gabion box structure is installed across the width of the wave flume as shown in Fig.4. Moreover, the 2 m width of the flume split along the middle of the flume with a 2 mm thick galvanised iron sheet for a distance of 18 m. It separates the wave flume into two parallel channels for the models to study. The first channel of the flume has a Geotextile tube structure type installed while the other has Geotextile tube with Gabion structure. The wall clearance between the model and either side of the flume wall as 2 cm. This configuration studied for various hydrodynamic coefficients and dissipation parameters under regular waves. Further, the two different cross-section of structures is shown in Fig.5 and 6. The performance of the structure for the design water depth of 0.5 m tested for a different range of wave period ranging from 1.5 sec to 4.7 sec under regular wave condition of varying heights.

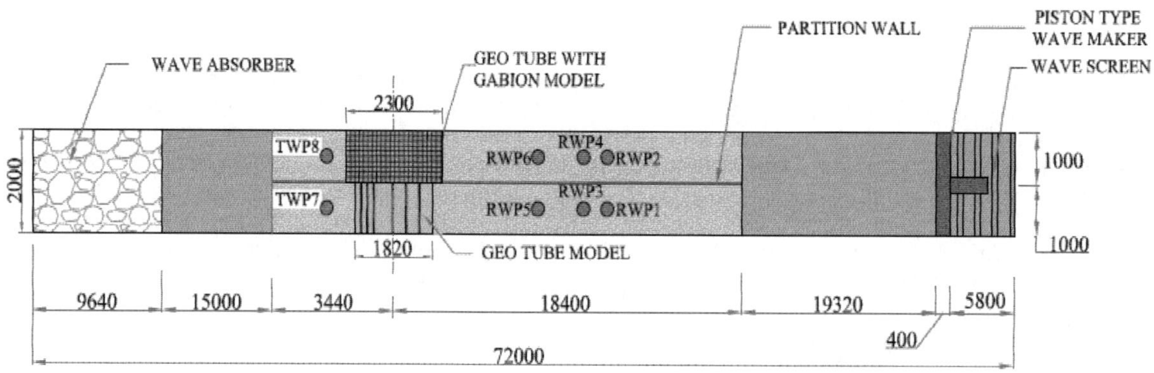

Fig. 3. Plan View of Wave flume with Models arrangement

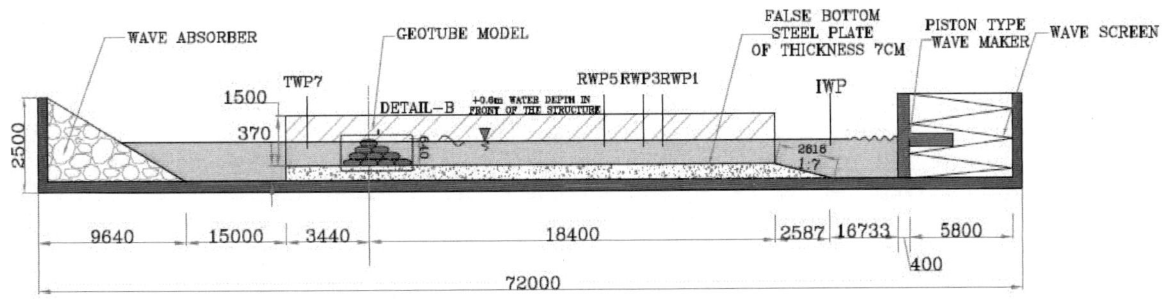

Fig. 4. Typical Cross Section of Wave flume with Models

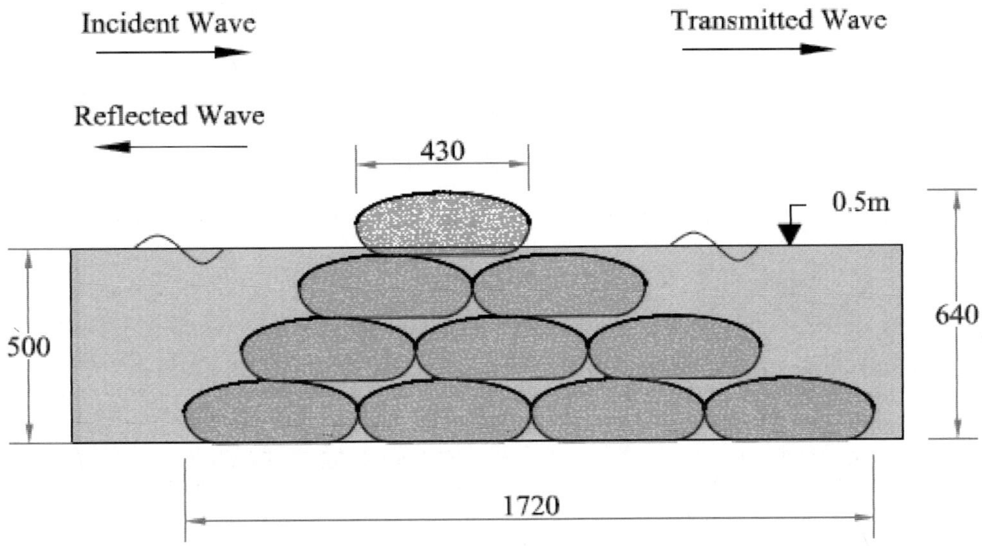

Fig. 5. Typical Cross Section of Geotextile tube Section (GTS)

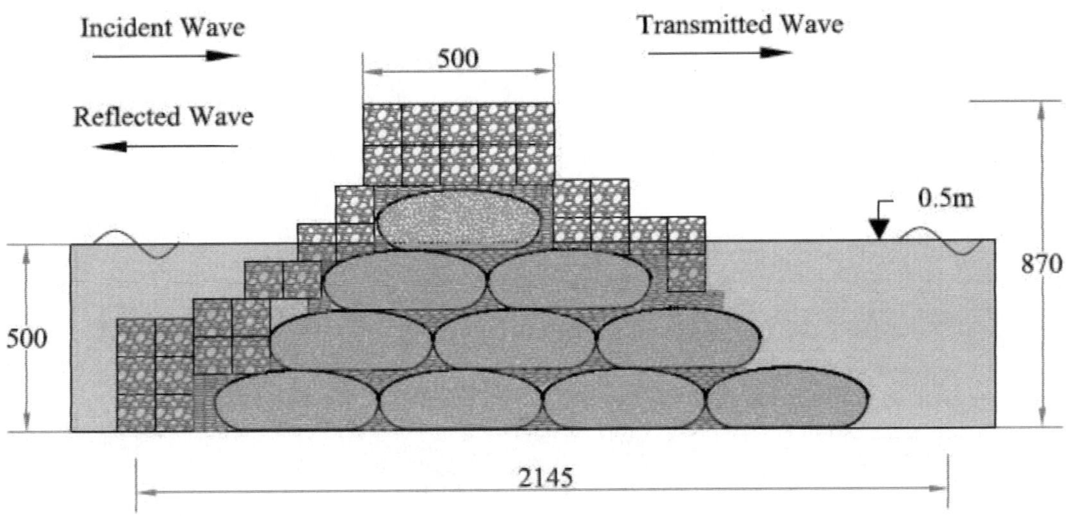

Fig. 6. Typical Cross Section of Geotextile tube with Gabion Section (GGTS)

3.4 Estimation of hydrodynamic coefficients

The effectiveness of the design in dissipating the incident wave energy is highly dependent upon the relationship between the wave characteristics, structural characteristics, and water depth. The hydrodynamic characteristics such as the reflection coefficient (K_R) and transmission coefficient (K_T) are obtained from the wave gauge measurements using three probe method (Mansard and Funke 1980). This approach provides the spectral energy of the incident, reflected and transmitted waves. To obtained the reflection and transmission coefficient, the losses (K_L) are calculated using Eq (1), by the conservation principle, i.e.

$$\boldsymbol{K_R^2} + \boldsymbol{K_T^2} + \boldsymbol{K_L^2} = \mathbf{1} \tag{1}$$

An attempt was made to examine the effect of reduction on the depth of submergence of the structure in attenuating the incident waves. The reduced depth of submergence is expected to reduce the cost of installation of the proposed structure while increasing the water exchange beneath the structure. Further such a measure can provide an insight into the hydrodynamic efficiency of the structure under extreme scenarios.

3.4.1 Reflection coefficient

Incident waves may be reflected (partially or wholly) from a beach and coastal or harbour structures, depending on the wave characteristics and the structure geometry. The magnitude of the reflection can represent by a reflection coefficient (K_R) as shown in Eq (2), which is nothing but the ratio of the reflected wave height (H_R) to the incident wave height (H_I). It can also obtain using wave energy as the square root, the ratio of the reflected wave energy (E_R) to the incident wave energy (E_I).

$$\boldsymbol{K_R} = \frac{\boldsymbol{H_R}}{\boldsymbol{H_I}} = \sqrt{\frac{\boldsymbol{E_R}}{\boldsymbol{E_I}}} \tag{2}$$

Impermeable vertical walls fully reflect and so that majority of the non-overtopping incident waves (i.e., $K_R \approx 1.0$). Beaches and sloped structures, however, reflect only a portion of incident wave energy. Several studies have been employed to estimate the amount of reflected energy regarding reflection coefficient (Harris and Sample 2009). Presently, the three-probe method used for determining reflection coefficient. It helps in the resolution of the incident and reflected amplitudes using least square technique and two-phase difference of the waves at three locations (Mansard and Funke 1980).

3.4.2 Transmission coefficient

The primary purpose of a breakwater or a coastal structure is to reduce the wave energy on its lee-side as well as to lessen the attenuation of approaching waves. The wave transmission is the wave energy which travels through a breakwater, either by passing through or by overtopping the structure. Wave energy attenuation in the lee-side of the breakwater is either dissipated by the structure (e.g., by friction, wave breaking, armour unit movement,) or reflected back as reflected wave energy (Yuliastuti and Hashim 2011). The effectiveness of a breakwater in attenuating wave energy measured by the amount of wave energy is transmitted or pass through the structure. Wave transmission is quantified by the using wave transmission coefficient by Eq (3).

$$\boldsymbol{K_T} = \frac{\boldsymbol{H_T}}{\boldsymbol{H_I}} = \sqrt{\frac{\boldsymbol{E_T}}{\boldsymbol{E_I}}} \tag{3}$$

Where, K_T is the wave transmission coefficient where, H_T is the height of the transmitted waves on the leeward side of the structure, and H_I is the height of the incident waves on the seaward of the structure. Alternatively, else regarding wave energy, one can rewrite as the square root the ratio of the transmitted wave energy (E_T) to the incident wave energy (E_I).

3.4.3 Loss Coefficient or Dissipation Coefficient

The portion of the energy judges the effectiveness of a coastal structure dissipated through friction, turbulence and wave breaking. Loss coefficient determined by the following relation given in Eq (4), loss Coefficient (K_L) is also called as Dissipation coefficient.

$$\boldsymbol{K_L} = \sqrt{(\mathbf{1} - \boldsymbol{K_R^2} - \boldsymbol{K_T^2})} \tag{4}$$

4 Results and discussions

The variation of K_R, K_T, $K_{L,}$ are studied for design water depth of 0.5 m. The former water-depth used for assessing the effect of high tides, whereas the latter one includes the high tide and storm surge. Results of the various hydrodynamic coefficients compared with non-dimensional parameter (D/L) for different wave steepness ranges H_{m0}/L, where, H_{m0} is the significant wave height obtained from the wave spectrum and D/L denotes the relative water depth. D is usually the water depth it crosses the structure from toe and L is the respective wavelength of the corresponding period of the regular wave were tested. In general, the present study confirms that the geotextile tube configurations structures have better hydrodynamic performance than the conventional rubble mound structures concerning reflection and dissipation coefficients. These results are discussed separately and compared for geotextile tube structure (GTS) and geotextile tube with gabion structure (GGTS). The variation of K_R, K_T, K_L with D/L for various wave steepness ratios are filtered and separated on three different wave steepness range, The results for K_R, K_T, and K_L are discussed for three different wave steepness range (H_{m0}/L) viz. Lowest wave steepness (0.001 to 0.01), Moderate wave steepness (0.01 to 0.02), and Highest wave steepness (0.02 to 0.038). Comparisons are discussed in the in the following sections.

4.1 Water depth (0.5m) to study high tide effects and storm surge

For the 0.5m water depth, the chosen water depth represents a scenario where the combined effects of high tide level and storm surge are studied. Herein, the water level is below 0.14 m for geotextile tube structure (GTS) beneath the crest of the structure height while the water level is 0.37 m beneath for geotextile tube with gabion structure (GGTS). The K_R and K_T are studied for the D/L range. Nevertheless, the K_L is in the range between 0.65 and 0.95 (Figure 6 (d-f)), meaning that the energy lost from the interactions of the waves due to the structure geometry, roughness, porosity effect of gabion boxes, wave breaking and run-up over the structure.

4.2.1 Influence of geotextile tube structure (GTS)

The GTS model (of 0.64 m height) is 0.14 m above the water depth. The effects of D/L range were 0.0483 to 0.165 on the hydrodynamic coefficients are studied in Figures 6 (a-c). For the H_{m0}/L range of 0.001 to 0.038 and maintaining D/L in the range of 0.0483 to 0.165, the K_R is decreasing from 80% to 30% as D/L increases and K_T is also reducing from 25% to 2%. K_L is increasing from 62% to 95%. For lower H_{m0}/L range of 0.001 to 0.01, the K_R is increasing rapidly from 50% to 80% due to long wavelength. Further, the K_T is found to be within a range of 2% to 27%. One must note that for lower H_{m0}/L values, $K_L$is increasing from 65% to 95%.

4.2.2 Influence of geotextile tube with gabion structure (GGTS)

For the GGTS D/L range is within 0.0483 to 0.165, were the freeboard height is 0.42 m above the water surface with a total height of structure as 0.92 m, since gabion boxes protect the geotextile tubes on the top. For the lower H_{m0}/L range i.e., 0.001 to 0.01, the hydrodynamic coefficients have a uniform change (Figures 6 (d-f)). In the given scenario, the K_R decreases from 75% to 50%, K_T increases from 2.5% to 30% and K_L increases from 67.5% to 99% as D/L increases. The physical reason for the increase in hydrodynamic coefficients in the lower wave steepness range is due to the influence of longer wavelength. The K_R is decreasing from 70% to 20%; the $K_T$is reducing from 15% to 2%, and K_L is increasing from 70% to 99%. For the smaller wave steepness (H_{m0}/L) range of 0.02 to 0.038, the K_R and K_T resemble lower values with a noticeable higher K_L value. It means that most of the energy is lost due to the interactions of the waves with the geotextile tube with gabion structure through wave breaking mechanism over the structure.

4.2.3 Comparison of geotextile tube structure (GTS) and geotextile tube with gabion structure (GGTS)

Comparing the different model cases (Figures 7 (a-c)) for estimating the efficiency of the geotextile tube with gabions, i.e., GTS and GGTS, the D/L is varied from 0.0483 to 0.165 for GTS while the D/L varies within 0.0483 to 0.165, In GTS, the K_R shows a large variation, *i.e.*, 0.37 to 0.8 for the lower H_{m0}/L range of 0.001 to 0.01; this is mainly due to the long period waves. Similar significant change is found for the transmission rate K_T, which implies loss coefficient, *i.e.*, the wave energy dissipation (due to porosity. Geometry, friction, *etc.*) has increased from 0.65 to 0.9. For GGTS, the K_R decreases while K_T increases along with wave steepness. Also, the K_L increases simultaneously which is due to the gabion boxes which are dissipating the wave energy. For both the cases, higher reflection and transmission coefficients are reported for longer wavelengths. Owing to the wave breaking from reflections, again the small, medium and high wave steepness cases are chosen based on visual observations. One can easily observe that; the hydrodynamic characteristics are better for the geotextile tube with gabion structure. As both the reflection and transmission coefficient for the GGTS with a freeboard of 0.37 m, may be the optimum height of the relatively submerged depth to dissipate the incident waves.

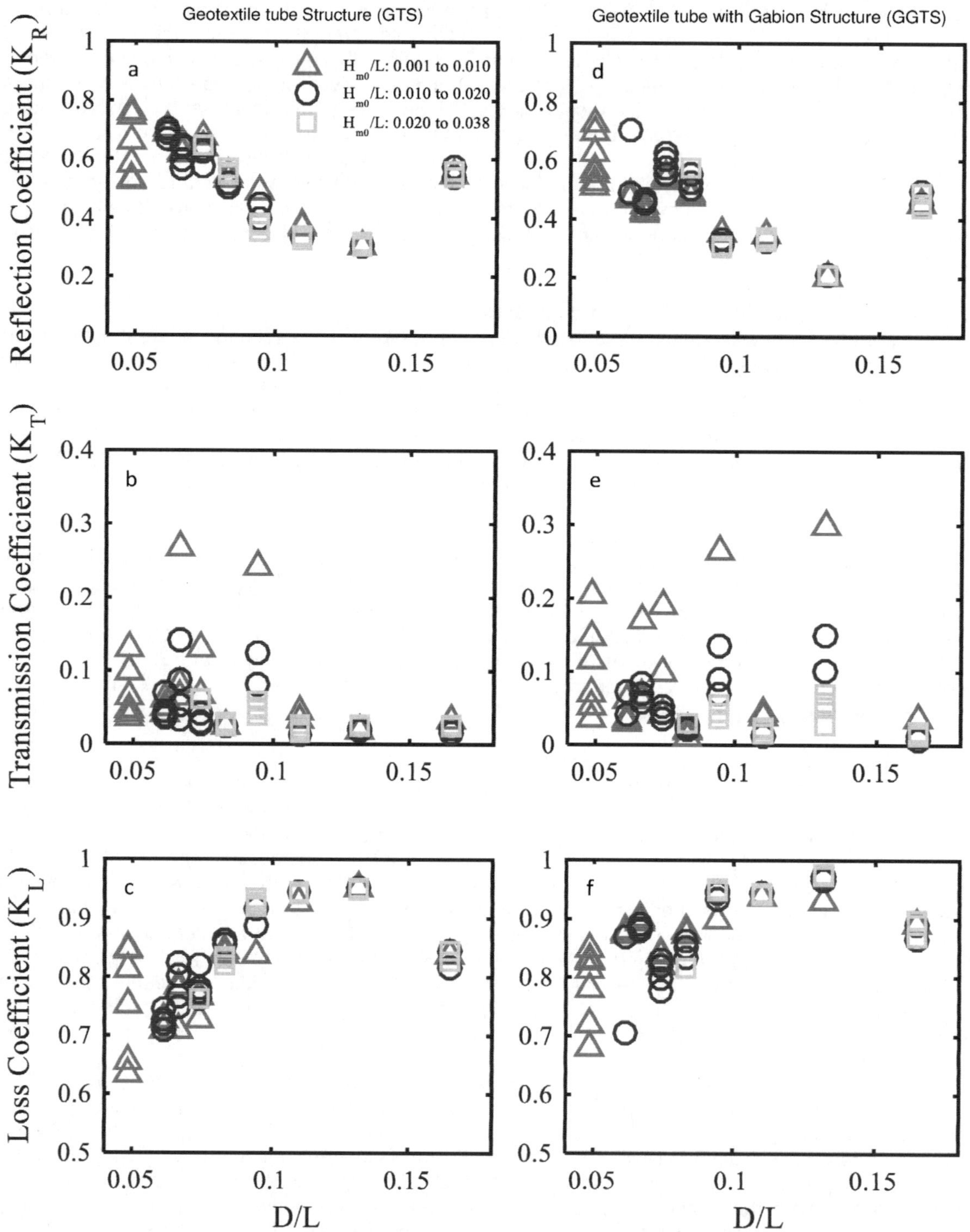

Fig.6 (a-c) Scatter plots of 0.5 m water depth showing a variation of *D/L* range over GTS. (d-f) Corresponding scatter plots for GGTS

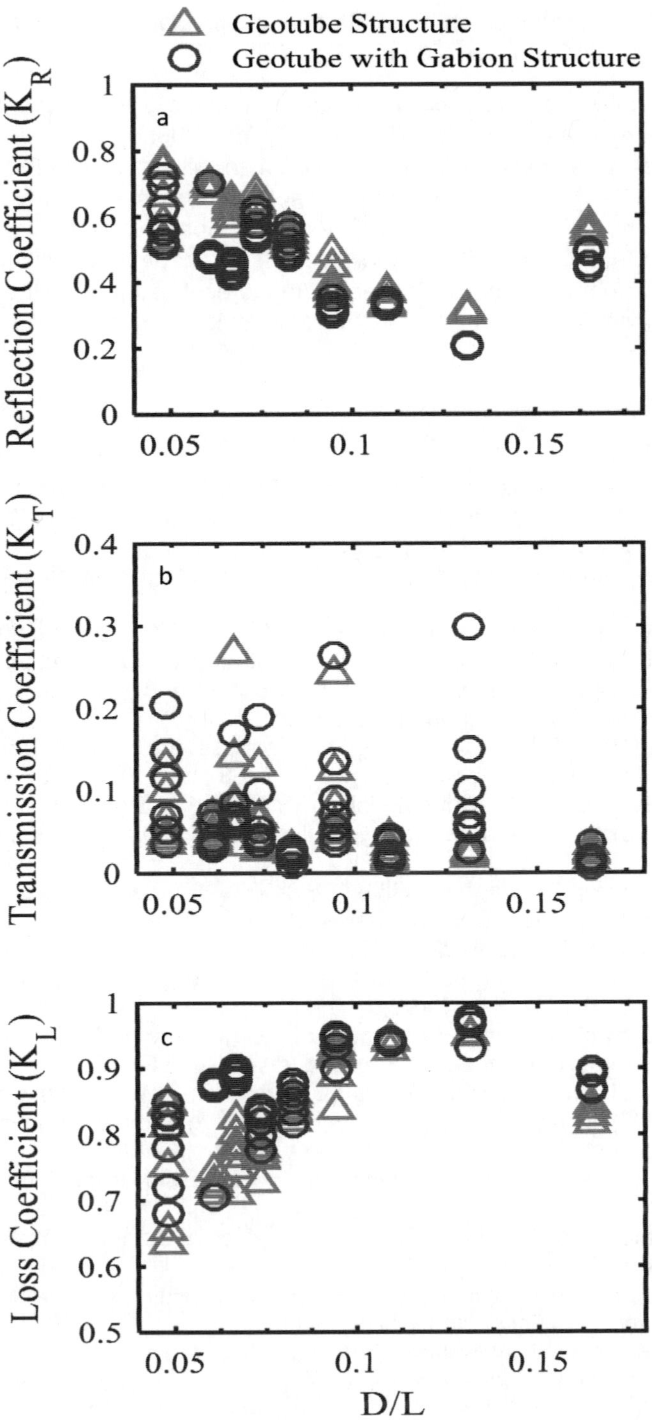

Fig.7 (a-c) Scatter plots of 0.5 m water depth showing *D/L vs* K_R, K_T, K_L for GTS & GGTS

5 Conclusions

The hydrodynamic performance of two different structure types has been examined and quantified. The wave reflection, transmission, and energy dissipation characteristics checked for regular waves of different wave heights and wave periods for design water depth. Both the models have higher energy dissipation characteristics. Usually, the reflection coefficient will be higher for long period waves.

However, Geotextile Tube with Gabion (GGTS) model provides a better reduction in reflection coefficients than the Geotextile Tube, section (GTS) model.

Considering the limitations of geotextile in coastal protection applications, a major problem with geotextile tube is that they have ultraviolet (U.V) stability even though it is eco-friendly to the marine environment. Geotextiles, when exposed to high UV radiation, will fail. Hence it is suggested to provide a model of Geotextile Tube with Gabion protection. It will preserve the integrity of the Geotextile tube core. Moreover, Gabion boxes are polypropylene tar coated mesh boxes filled with a smaller range of stones which increases that porosity. Such Gabion boxes are capable of dissipating large kinematic wave forces than the conventional monolithic coastal structures. In addition to these measures, Gabion boxes protect the Geotextile Tubes from various hydrodynamic and geotechnical failure.

6 References

Bezuijen A, De Groot MB, Breteler MK, Berendsen E (2004) Placing Accuracy and Stability of Geocontainers. In: Proceedings 3rd European Geosynthetics Conference EuroGeo3, Munich

Cho S-M (2009) Geotechinical design of the Incheon Bridge. Korea

D'Angremond, Berg KJF van den, Jager JH de (1992) Use and Behaviour of Gabions in Coastal Protection. Coast Eng 1748–1757

Das Neves L, Lopes ML, Veloso-Gomes∞ F, Taveira-Pinto F (2009) Experimental Stability Analysis of Geotextile Sand-filled Containers for Dune Erosion Control. J Coast Res 56:487–490

Harris LE, Sample JW (2009) The Evolution of Multi-Celled Sand-Filled Geosynthetic Systems for Coastal Protection and Surfing Enhancement. Reef J 1:1–15

Heerten G, Wittmann L (1985) Filtration properties of geotextile and mineral filters related to river and canal bank protection. Geotext Geomembranes 2:47–63 . doi: 10.1016/0266-1144(85)90011-1

Jackson A (2006) Failure Modes & Stability Modelling for Design of Sfgs. 30th Int Conf Coast Manag

Lawson CR (2008) Geotextile containment for hydraulic and environmental engineering. Geosynth Int 15:384–427 . doi: 10.1680/gein.2008.15.6.384

Leshchinsky D, Leshchinsky O, Ling HI, Gilbert P a. (1996) Geosynthetic Tubes for Confining Pressurized Slurry: Some Design Aspects. J Geotech Eng 122:682–690 . doi: 10.1061/(ASCE)0733-9410(1996)122:8(682)

Luettich SM, Giroud JP, Bachus RC (1992) Geotextile Filter Design Guide. Geotext Geomembranes 11:355–370 . doi: 10.1016/0266-1144(92)90019-7

Mansard EPD, Funke ER (1980) The Measurement of Incident and Reflected Spectra Using a Least Squares Method. Proc 17th Conf Coast Eng Sydney, Aust 154–172

Pilarczyk K (2008) Alternatives for coastal protection. J Water Resour Enviromental Eng 181–188

Shin EC, Oh YI (2007) Coastal erosion prevention by geotextile tube technology. Geotext Geomembranes 25:264–277 . doi: 10.1016/j.geotexmem.2007.02.003

Takahashi S (1997) Breakwater Design. In: Handbook of Port and Harbor Engineering. Springer US, Boston, MA, pp 951–1043

U. S. Army Engineer Watetways Experiment Station (1986) Use of Gabions in the Coastal Environment. Coast Eng Tech Note 3:

Van Steeg P, Vastenburg E, Bezuijen A, et al Large-Scale Physical Model Test on Sand-Filled Geotextile Tubes and Containers Under Wave Attack

Welsby JMM (1984) A Review of Novel Shore Protection Methods

Wuthrich D, Chanson H (2014) Hydraulics, Air Entrainment, and Energy Dissipation on a Gabion Stepped Weir. J Hydraul Eng 2:1–10 . doi: 10.1061/(ASCE)HY.1943-7900.0000919.

Yuliastuti DI, Hashim AM (2011) Wave Transmission on Submerged Rubble Mound Breakwater Using L-Blocks. In: International Conference on Environmental Science and Technology. pp 243–248

AUTHOR INDEX